MULTIPLE CROPPING AND TROPICAL FARMING SYSTEMS

T0230953

THIS BOOK
IS FOR THOSE WHO ARE
WILLING AND ABLE TO HELP
THE TROPICAL SUBSISTENCE FARMER
IMPROVE HIS LOT

Multiple Cropping and Tropical Farming Systems

WILLEM C. BEETS
The Asian Development Bank,
Manila

CRC Press
Taylor & Francis Group
Boca Raton   London   New York

CRC Press is an imprint of the
Taylor & Francis Group, an **informa** business

First published 1982 by Westview Press

Published 2018 by CRC Press
Taylor & Francis Group
6000 Broken Sound Parkway NW, Suite 300
Boca Raton, FL 33487-2742

CRC Press is an imprint of the Taylor & Francis Group, an informa business

Visit the Taylor & Francis Web site at
http://www.taylorandfrancis.com

and the CRC Press Web site at
http://www.crcpress.com

British Library Cataloguing in Publication Data

Beets, Willem C.
    Multiple cropping and tropical farming systems.
    1. Crop yields
    I. Title
    338.1'6   SB186

ISBN 13: 978-0-367-00663-1 (hbk)
ISBN 13: 978-0-367-15650-3 (pbk)

Contents

Figures

Preface

Due to growing concern over the world food situation, multiple cropping is receiving more and more attention in the developing world as well as in some developed countries. It is now recognized that the introduction of well-planned multiple cropping practices is one of the more feasible ways of raising agricultural production in the tropics.

As defined in this book, multiple cropping means growing more than one crop on the same piece of land during one calendar year. It is not a new concept. Rather, it is an age-old method of intensive farming, found throughout the tropics, by which land-use and labour productivity are maximized.

In the context of greatly increased interest in higher food production, stemming from recent food shortages and bleak production forecasts, multiple cropping has a strong link with two particularly significant phenomena which have attracted a great deal of public attention in the last decade - the "Green Revolution" and the "Energy Crisis."

The Green Revolution involving rice and wheat, is commonly associated with the introduction of high yielding varieties, and not with other aspects of the production system. However, one important lesson agriculturalists have learned in the past decade is that the improvement of one production factor does not, by itself, lead to higher output; the major impact of the new varieties on production occurs only when cultural practices and cropping systems are improved simultaneously. Substantial production increases can, therefore, be expected when the new varieties are used in multiple cropping systems, especially when better overall resources and support services are available.

Multiple cropping, in essence, represents a philosophy of maximum crop production per unit of land by producing several crops within one calendar year, maximizing use of available solar energy and other natural resources. Solar energy is abundant in the tropics, while fossil energy is usually scarce. Multiple cropping, therefore, seems most appropriate with the present shortages of fossil energy.

Multiple cropping and farming systems are complex; currently only scanty research data are available and comprehensive studies are called for. However, comprehensive studies of all parameters involved are not yet feasible because it would be unyielding. But, at the minimum, an intuitive knowledge of interacting factors is essential and a philosophical, rather than an empirical, approach has often been used. This is particularly true in the discussion of socio-economic aspects of whole multiple cropping systems.

Many questions have been raised in this book, but not all have been answered. On occasion, I felt it was sufficient to raise a question and bring it to the attention of the reader. At other times, the answers are still not available, but it seemed worthwhile, nevertheless, to touch upon the matter. My approach has been multidisciplinary in order to try to make the book attractive to a broad readership. Specialists may feel that I have sometimes only skimmed the surface of the topic, but this was a deliberate strategy, and I hope that this book will also prove useful to the specialist, since it explains how his particular speciality fits into the broad and complex scheme of tropical farming systems.

I accept all responsibility for errors and omissions and welcome suggestions for improvements.

Willem C. Beets

I Introduction

TROPICAL FARMING SYSTEMS IN GENERAL

The many forms of agriculture found throughout the world are the result of variations in local climate, soil, economics, social structure and history. Water balance, radiation, temperature and soil conditions are the main determinants of the physical ability of crops to grow and farming systems to exist. Human factors that play dominant roles include social, economic and political considerations such as tradition and religious convictions; prices and ease of transport; the existence of marketing channels; stability of prices and availability of capital and credit.

Farming systems also depend heavily on the character of production, i.e., whether the crops are produced in a subsistence or a commercial economy. One of the main features of subsistence farming is that the farmer has to produce crops in order to live. Consequently, he often resists changing production methods since, when the changes turn out to be unproductive, his livelihood and survival are threatened.

The way crops are grown further depends on the level of technology and the land area available. At high levels of technology and land abundance there is generally a high level of mechanization, and uniformity of land, soil fertility and genotype are needed. On the other hand when land is scarce, cropping systems tend to be more intensive and less mechanized.

When the above broad factors are taken into consideration, the main specific determinants of farming systems can be summarized as follows:

    (i)   land availability and population density;
    (ii)   type of crop rotation;
    (iii) water supply;
    (iv)   cropping pattern;
    (v)   type of implements used for cultivation; and
    (vi)   degree of commercialization.

And by using these determinants the following three main tropical farming systems can now be recognized:

    (i)   extensive shifting cultivation;
    (ii)   intensive subsistence agriculture; and
    (iii) commercial, frequently mechanized, crop tillage.

1

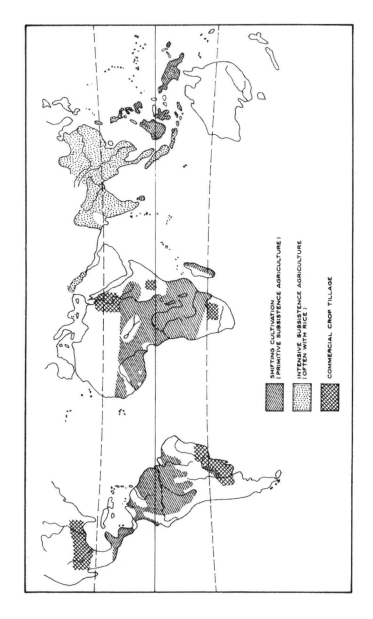

Figure 1.1 Global distribution of the three main tropical farming systems. (After Whittlesey, 1974; Wagner and Mikesell, 1965).

The legend for the map reads:

SHIFTING CULTIVATION (PRIMITIVE SUBSISTENCE AGRICULTURE)

INTENSIVE SUBSISTENCE AGRICULTURE (OFTEN WITH RICE)

COMMERCIAL CROP TILLAGE

The distribution of these farming systems is shown in Figure 1.1. Multiple cropping, defined as growing more than one crop on the same piece of land during one calendar year, takes place in different forms in all three farming systems.

For the purpose of this book the following typology of the main multiple cropping systems has been adopted:

(i) Mixed cropping is defined as growing more than one crop on the same piece of land at the same time. It is common in shifting cultivation systems which occur in almost half the tropical world. Intercropping is a form of mixed cropping where all crops are planted in a fixed pattern of spacings and rows.

(ii) Relay cropping is defined as planting crops between plants or rows of an already established crop during the growing period of the first planted crop(s). It is widely practised in intensive subsistence agriculture in areas such as Asia, China and South America; and

(iii) Sequential cropping is defined as growing more than one crop on the same piece of land with each crop during a different time of the year. It is common in Asia and China with intensive subsistence agriculture. Double and triple cropping are common forms of sequential cropping.

The multiple cropping systems mentioned above can be described and classified in more detail using the following criteria:

(i) The degree of intensification in space, or, the level of intimacy of the crop species;

(ii) The degree of intensification in time, or, the crop intensity over the year; and

(iii) The relative time of planting of crop species.

Mixed cropping has the highest level of intimacy (the different species are planted close to each other) and sequential cropping has the lowest (also see Fig. 1.2). The intensity of the system depends primarily on the degree of intensification in time, e.g., triple cropping (three successive crops in one year) is more intensive than one mixed crop of two species per year since the mixed cropping system only uses about four months of the year, whereas in triple cropping the land is covered with a crop during almost the entire year.

The potential productivity of a multiple cropping system can be described by using the concept of the Multiple Cropping Index (MCI) which is given as:

3

$$MCI \; = \; \frac{CROP \; AREA \; FOR \; ONE \; YEAR}{CULTIVATED \; AREA \; FOR \; ONE \; YEAR} \; x \; 100 \; per \; cent$$

A high MCI means intensive land use and high annual yield potential. Whether this potential is utilized depends on the productivity of the individual crops or species in the multiple cropping system. In the tropics, the yield of individual crops is usually far below the potential, and yields can often be increased with relatively simple changes in production methods.

INCREASING THE PRODUCTIVITY OF TROPICAL CROPPING SYSTEMS

Crop production can be increased by one or more of the following:

(i) by expanding the area planted to crops;
(ii) by raising the yield per unit area of individual crops; and
(iii) by growing more crops per year (in time and/or in space).

In the past, agricultural production has been mainly increased by (i) cultivating more land, but now there is limited scope for this since unused land is rapidly diminishing. More recently there has been greater emphasis on (ii) increased yield per unit area. This has been especially so in the more developed, temperate countries. In the developing countries in the tropics emphasis has often been on (iii) growing more crops per year, or multiple cropping. Theoretically, the highest possible production would be achieved by using all three possibilities, i.e. by continuously growing high yielding crops on the maximum land area available.

Crop production is a complex process and in practice there are always constraints to the adoption of new practices which achieve high yields. These complexities and the constraints resulting from them can best be understood if one considers crop production to be the result of two multidimensional vectors, the environment (E) and the plant genotype (G). The crop yield (y) is the result of the interaction of the two vectors E and G:

$$y = f \; (E, \; G)$$

The Genotype is the aggregation of individual plants, frequently of similar constitution, grown in a particular location for a specific product required by man; and

4

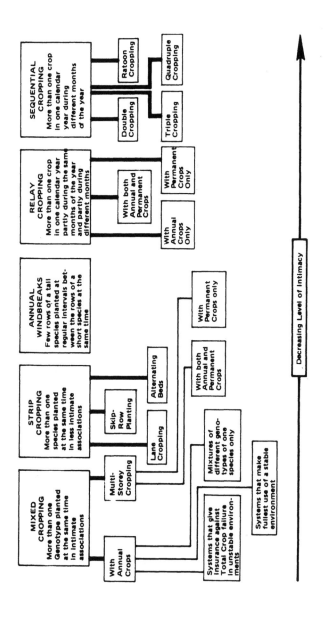

Figure 1.2   Schematic presentation of the definitions of the principal multiple cropping systems.

The Environment includes all micro-climatological and physical factors such as water, radiation, temperature, evaporation and soil conditions as well as human, management, economic and political considerations.

Using this concept, Figure 1.3 gives a diagrammatic summary of the ways through which the yield of individual crops can be increased. The ways listed in Group I are all agronomic or crop husbandry techniques and all require little capital investment. Those in Group II are rather more demanding - some capital is required and the changes can only be brought about using the services of technical specialists. Changes listed in Group III are the most difficult to bring about but have frequently been implemented in the developed countries where technical know-how and capital are not major constraints.

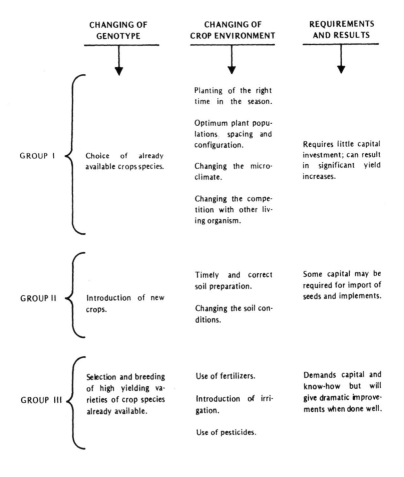

Figure 1.3. Ways to augment the yield of individual crops.

The foregoing refers to yield improvement of individual crops. When methods of augmenting the productivity of whole cropping systems over a whole year are considered the productivity of single crops is only one factor that determines the output of a given piece of land. In multiple cropping systems it is important to consider the interactions between different crops or species - the effect one crop has on another is important. In mixed cropping systems, one of the species in the crop association normally gives a lower yield per unit area than when it is grown as monoculture. However, the combined yield of the two species is higher than the yield of the sole crop. Similarly, when only one rice crop is grown per year, this crop can be grown at the optimum time of the year, a high yielding late maturing variety can be used and the sole crop can give maximum yield. In sequential cropping, one of the crops is generally grown during a period which has sub-optimum growth conditions and the yield of the individual crops grown in sequence will be lower than when one crop per year is grown. However, the combined yield of the two sequential crops is higher than that of the single rice crop. In summary, in multiple cropping, the total output of a given land area is more important than maximum yields of single crops.

REASONS FOR THE ADOPTION OF MULTIPLE CROPPING SYSTEMS

The main reasons for adopting multiple cropping systems can be classified into two broad groups: (a) physio-technical, and (b) socio-economic. In practice there is a considerable interdependence between these groups.

The physio-technical reasons fall into three subdivisions:

(i) Better utilization of environmental factors: Plants of different growth habits often have different environmental requirements. When crops are grown in mixtures for a given area and time, the utilization of light is maximized since the plant canopies of the two or more crops can together intercept and utilize more light. Crops with different rooting habits may together be able to take up more nutrients than one crop.

Sequential cropping systems make better use of land and solar energy since these systems occupy the land during more months of the year which means more photosynthetic opportunity and greater nutrient and water-use resulting in higher annual yield;

(ii) Greater yield stability in variable environments: Environmental variability usually results in yield instability. When crops are grown in associations, this yield variability is often reduced because the different species are not equally affected by an adverse environment. For example, consider growing a mixture of maize and upland rice in successive "dry" and "wet" years. In the wet year the rice will do well, but the maize will give a low yield because of excessive moisture. In the dry year the maize will produce well but the rice will suffer from moisture stress.

Yield instability in the tropics is often caused by pests and diseases. When crops are grown in mixtures, a serious outbreak of a certain pathogen usually attacks only one of the species in the crop association, resulting in a yield decrease of this species but not of the other species, the yield of which may even increase; and

(iii)   Soil Protection:
When crops overlap in terms of the time they are in the ground, the period of the year during which the ground is protected by leaf-cover is extended, reducing the physical damage by rain, wind and soil erosion. All multiple cropping systems provide better soil cover than sole crops and multiple cropping is therefore highly desirable on unstable soils.

The socio-economic reasons for the adoption of multiple cropping systems fall into two subdivisions:

(i)   Magnitude of inputs and outputs:
Generally, a higher yield and greater gross return per unit area can be obtained with multiple cropping. The principal extra input to achieve this higher output is "labour". In many tropical farming systems, labour cannot be seen as an "input" in economic terms since the opportunity cost for labour is very small. In subsistence agriculture the return from the farmers' effort is more important than the amount of effort or labour required; and

(ii)   Regularity of food supply:
When planting and harvesting is done in phases, and several crops instead of a few are grown, a regular and varied supply of food for the household is assured and storage losses are minimized.

The adoption of multiple cropping in tropical countries where capital is a limiting factor has several advantages. For example, scarce external inputs such as fertilizers, pesticides, fuel, etc. should be used to the fullest degree. There is generally less wastage of these resources

8

in multiple cropping systems than in sole cropping since there is less leaching of fertilizer when more than one species is grown, and less land preparation is required. The energy crisis has retarded the agricultural development of many tropical countries as it has made fertilizers and other energy-intensive agricultural inputs even more scarce and expensive than they were previously. Therefore, it is now of paramount importance to adopt agricultural practices and cropping systems that make the best possible use of these inputs.

CONSTRAINTS ON THE DEVELOPMENT OF MULTIPLE CROPPING PRACTICES

In many countries multiple cropping practices, and especially the most common forms of multiple cropping (mixed and relay cropping), are often associated with backward, peasant type farming. Frequently, the first thing agronomists and agricultural extension workers did, when they tried to improve traditional agriculture, was to advise against mixed cropping. They argued that improvements are only made when single crops are planted in rows. The reason for this seems to be that research on multiple cropping practices in the tropics has for long been neglected as a result of the influence of western research which has been biased towards monocultures. Mechanization has played an important role in the development of agriculture in the developed countries and, because of it, cultural practices, varieties, harvesting techniques, etc. have had to be adapted. Important prerequisites for mechanization were, and often still are, monocultures, row-culture and uniformity of crops.

After the Second World War, many attempts were made to introduce Western techniques into the tropical developing world. In many cases these attempts have led to enormous failures -the Tanzania groundnut scheme serves as a classic example. Agriculturalists with a Western background or education often find it difficult to visualize multiple cropping systems under improved technological conditions.

The problems usually articulated include:

   (i)   research problems on improved varieties and cultural practices such as weed control, fertilization, insect control, etc., are compounded when dealing with more than one crop; and

  (ii)   it is difficult to visualize the successful introduction of farm mechanization into systems that are dominated by crop mixtures.

The problems noted are real, but in most cases satisfactory solutions can be found. However, for the average tropical farmer at the present time, and for the forseeable future, both are irrelevant considerations. He is not

9

concerned with research or with efficiency and optimum yields - his prime concern is to assure a sufficiently high yield to feed his family. What is very important to him is that he is certain that he will have some return from what he has planted. The growing of crops in mixtures remains a basic characteristic of farming under present conditions and this book attempts to demonstrate that there are valid reasons of a technological and socio-economic nature for farmers' reluctance to change to sole-cropping systems.

Until recently, researchers have been hesitant to tackle mixed cropping experiments because of the large number of crop combinations and factors that interact. Fortunately, there has lately been an appreciation by some research workers that certain multiple cropping systems have great potential and that the problems associated with such systems should not prevent research on multiple cropping under improved technological conditions being undertaken.

Most tropical farmers still remain unconvinced of the value of very drastic changes in their farming methods, e.g. from mixed cropping to sole cropping and from broadcasted seed to row-culture. Until agricultural scientists in the developing world can suggest modifications that have a convincing return and yet do not involve large changes in existing farming methods, it is unlikely that they will be successful in improving traditional agriculture.

GEOGRAPHICAL AREAS WHERE MULTIPLE CROPPING CAN BE FOUND

Dalrymple (1971) surveyed the occurance of multiple cropping systems throughout the tropics, and concluded that multiple cropping is a widespread practice. It is estimated that 98 per cent of cowpeas, probably the most important legume in Africa, is grown in association with other crops (Arnon, 1972). Norman's survey in Northern Nigeria (1975) reports mixed cropping on 83 per cent of all cropped land. In Columbia 90 per cent of the bean crop is grown in association with maize, potato and other crops, while in Guatemala 73 per cent of bean production is from mixed cropping. Frances and Flor (1975) estimate that in the Latin American tropics, 60 per cent of the maize is associated with other crops.

Furthermore, in Asia and China, there are only few areas where the Multiple Cropping Index is less than 150. Usually, all land is planted with rice at least once a year and after the rice crop is harvested, a second crop such as soya beans, mung beans or maize is grown. Figure 1.4 shows where some multiple cropping systems occur and it clearly illustrates that multiple cropping is a widespread practice.

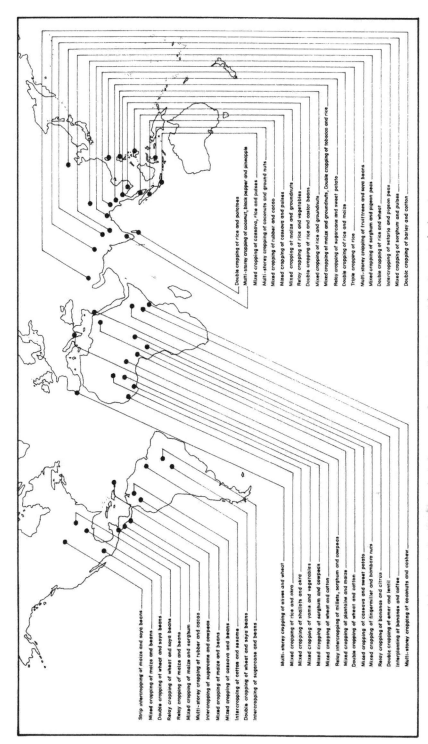

Figure 1.4 Some multiple cropping systems practised or tried out.

11

Multiple cropping systems and climate

Year-round adequate temperature and solar radiation (roughly over 200 cal/sq cm/day) are the main prerequisites for
multiple cropping. These conditions normally prevail in
the tropics (with the exception of minor areas at high
altitude) and in most of the subtropics. The next consideration is the availability of water, either through rainfall or irrigation. The distribution of rainfall varies
greatly between areas of comparable temperature and the
annual moisture balance is, therefore, the main factor that
determines the type of cropping system to be found in particular areas. In areas with a uni-modal rainfall pattern
of relative short duration (3-5 months), the moisture
balance only allows for crop growth during 4-6 months of
the year (without irrigation) and it is therefore important
that maximum use is made of the (moisture) season. Generally one crop cannot fully exploit such a short moisture
season. Mixed cropping or a combination of mixed and relay
cropping is better able to exploit the environment because
the two or more species will together have a longer leaf
area duration, and will extract more nutrients and water
than a single crop. Short moisture seasons are common in
Africa, and mixed and relay cropping systems are therefore
very popular on this continent. In Latin America and Asia,
both uni-modal and bi-modal rainfall patterns prevail and
the moisture seasons in both continents (with the exception
of India) are generally longer than in Africa. For these
areas moisture is less of a constraint. In this situation
sequential cropping or a combination of sequential, relay,
and mixed cropping can best exploit the environment. Multiple cropping in Asia is often built around a wet season
crop of rice. During the dry season the land may be planted again to rice but it is often devoted to leguminous
crops.

Cropping systems and soil

The soil requirements for multiple cropping are basically
the same as for other forms of intensive crop production.
When the soil is infertile, a crop association with different rooting habits can often assure a reasonable production where sole crops give only marginal yields because the different species have together more access to
the limited nutrients.

In the design of cropping systems, plant population and
crop intensity are often determined by the soil fertility -
the higher the soil fertility, the more plants or crops are
required to exploit the environment. When the natural soil
fertility is low, and no fertilizers are available, sequential cropping is not generally desirable, but mixed cropping can be advantageous. The latter situation is quite
common in tropical rain forest areas in the wet equatorial
region, especially in Latin America, where soils are often
relatively infertile due to leaching. In these regions

multiple cropping can also be advantageous since a better soil cover throughout the year will protect the soil from rain damage and erosion (also see Chapter V). When, in a humid tropical region, the soil is fertile, relay and sequential cropping systems have high yield potentials.

Cropping systems and population

Population pressure is an important determinant of farming systems, especially of shifting cultivation systems. As population pressures increase in shifting cultivation areas, the cropping system has to change because there is no longer enough land available to allow for the long fallow periods which are part of these systems. Multiple cropping systems are often, therefore, the only way of providing a livelihood for the increased population.

In areas of intensive subsistence agriculture and high population pressures, labour is normally abundantly available and multiple cropping is the logical way to produce crops. The systems are more productive, and able to feed a larger population and, at the same time, reduce unemployment (also see Chapter IV). This is illustrated by the fact that countries with high population densities, such as Taiwan and India, are invariably countries with high multiple cropping indices.

II The different multiple cropping systems

INTRODUCTION

This Chapter discusses the different multiple cropping systems defined in Figure 1.2 and gives examples for each system. When classifying the systems it should be recognized that many multiple cropping systems practised by farmers are in fact combinations of different systems. Combinations of mixed and relay cropping are particularly common.

Classifying the different systems is sometimes difficult because there can be a gradual transition from one system to another. For example, when the spatial arrangement of mixed cropping is changed, it can become strip cropping, which in turn, can become an annual windbreak system. The different systems are discussed in order of decreasing level of intimacy, starting with mixed cropping which is the most intimate system.

MIXED CROPPING WITH ANNUAL CROPS

Mixed cropping is defined as growing more than one species on the same piece of land at the same time, or with a short interval. The different species are either mixed in an organized manner, with a fixed pattern of spacings and plant populations, or, in an unorganized manner, where species are unevenly distributed over the land. The latter is common in subsistence agricultural all over the world. Figure 2.1 shows a typical mixed cropping system in African subsistence farming. In mixed cropping, there is no distinct row arrangement. Row intercropping is a form of mixed cropping where all crops are planted in a fixed pattern of spacing and rows. The latter is illustrated in Figures 2.2 and 2.3.

Mixed cropping is practised for various reasons. In subsistence agriculture, especially where there is a highly variable and unstable environment, it is an insurance against total crop failure. In more stable, favourable environments, higher total yields can be obtained per unit of land because the available resources such as light, nutrients and water are better utilized than in sole cropping.

Figure 2.1 Mixed cropping system in African subsistence farming.

One of the oldest problems in crop production is the inability to forecast the weather. Most crops need certain weather conditions to be able to fully exploit the environment. When these conditions do not prevail, it is likely that another crop species with environmental requirements which fit the weather conditions better during that particular season would give higher yields. Hence, in some years, crop A or species A will succeed, and in others crop B will do better. Since, at the time of planting the farmer does not know what weather conditions are going to occur during the growing period, both crops are planted in a mixture. Thus, when growing conditions turn out to be unsuitable for crop A, crop B might still produce thereby avoiding complete crop failure. It is, of course, possible to divide the available land and plant both crops in separate areas. However, planting the total area available with a mixture of the two crops is a better proposition since in the former case, the land devoted to the crop that totally fails will be completely unproductive, whereas, in the latter case, part of the resources of the area initially devoted to the species that fails can to some extent be used by the other species in the mixture. Consequently, the species will produce more per square meter when mixed cropped than the same number of plants will produce when half the amount of land is sole cropped. Andrews (1973) calls this a "safety factor" while other authors call it a "risk factor" or "security factor" (IRRI, 1975).

From the above it can be concluded that for mixed cropping to be advantageous, the components of the crop association should have different environmental requirements which generally means contrasting habits. Therefore, crop combinations which are very common include maize and soya beans; rice and pulses; maize and bambara nuts (Voandseia subterranea); and maize and cowpeas. Note that all combinations are associations of a legume and a cereal.

Crop mixtures that make fullest use of a stable environment

When the environment is less variable and genotypes are well adapted there are also advantages for mixed cropping since frequently no one species on its own can fully exploit the environment. If the soil fertility is controlled and at an optimum level and if the moisture balance is also controlled, possibly by means of irrigation, light often becomes a limiting factor in plant production. In this situation, light in sole cropping systems will either be wasted in the early part of the growing season when the individual plants are small or there will not be enough light later in the season when plants overshade each other. When two species of contrasting habits are grown in association, their light requirements are somewhat spread over the growing season. This will result in higher total

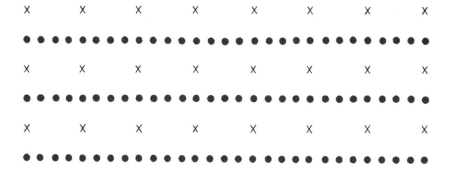

Figure 2.2   Diagrammatic presentation of the spatial
             arrangements of a row intercropping system
             of maize (x) soya beans (•).

Figure 2.3   Row intercropping system of sorghum and soya
             beans on an experimental farm in Zimbabwe.

light interception over the entire growing period. As a second example, consider two species of contrasting rooting habits. One species frequently takes up large quantities of one particular soil nutrient while the other extracts large quantities of another nutrient. In these examples, mixed cropping will result in better light use or higher total nutrient uptake and therefore, in greater total growth and yield.

Mixtures of different genotypes of the same species

Extensive selection and breeding of crops has frequently led to great uniformity of genotypes such as in single-cross hybrids of maize. Genetic variability in such types is very small and the variety is therefore suitable only to a narrow range of environments. Consequently, in a variable environment, sometimes one genotype cannot fully exploit it and it may, therefore, be desirable to plant a mixture of different genotypes.

Another reason why multi-strain varieties may be superior is that a mixture has a greater tolerance to diseases and pests (Schwerdfeger, 1954). When a stand of plants that are susceptible to a disease is "diluted" with resistant plants, the level of infestation of, or damage to, individual susceptible plants may be reduced. When fewer plants are attacked, the overall effects of disease and pests are reduced.

RELAY CROPPING WITH ANNUAL CROPS

Relay cropping is defined as planting crops between plants or rows of an already established crop during the growing period of first planted crop(s). The interplanted young plants not only gain more time for growth on the same piece of land, but can also make use of the residual fertility of the previous crop and the remaining moisture in the soil. The spatial and time arrangement of this form of multiple cropping is illustrated in Figure 2.4. Maize is planted at the onset of the rains and cassava is planted later in the season. The cassava completes its growth cycle at the end of the rainy season by using residual moisture, thereby using the resources to the fullest. This form of relay cropping is widely practised in Indonesia and is illustrated in Figure 2.5.

Time of planting and harvesting and length of overlap periods are critical in relay cropping systems. Consequently, it is often necessary to breed varieties specially suitable for relay cropping practises - the main characteristic being earliness of maturing.

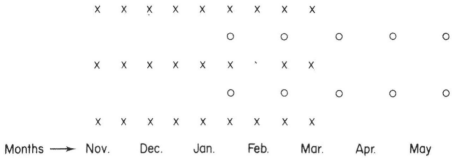

| Months ⟶ | Nov. | Dec. | Jan. | Feb. | Mar. | Apr. | May |

Figure 2.4 Diagrammatic presentation of a relay-cropping system. Maize (x) is planted at the beginning of the growing season (November) and cassava (o) is later interplanted.

Figure 2.5 A very intensive multiple cropping system in Indonesia. Maize is planted at the beginning of the growing period; cassava is later relay-interplanted and cowpeas are mixed with the cassava. Maize plants are topped to provide forage for livestock.

An interesting multiple cropping system from Nigeria which has an element of relay cropping is described by Andrews (1975). Three types of crops were used - a quick maturing cereal (Pennisetum millet or maize), a long season cereal (Guinea zone dwarf sorghum) and a quick maturing, though late-planted legume (cowpeas). Figure 2.6 illustrates how the crops were planted.

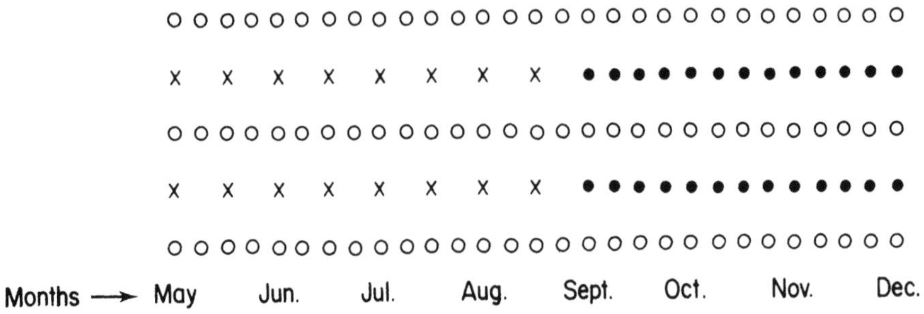

```
o o o o o o o o o o o o o o o o o o o o o o o o o o o o o o o o

x  x  x  x  x  x  x  x  ● ● ● ● ● ● ● ● ● ● ● ●

o o o o o o o o o o o o o o o o o o o o o o o o o o o o o o o o

x  x  x  x  x  x  x  x  ● ● ● ● ● ● ● ● ● ● ● ●

o o o o o o o o o o o o o o o o o o o o o o o o o o o o o o o o
```

Months ──► May   Jun.   Jul.   Aug.   Sept.   Oct.   Nov.   Dec.

Figure 2.6   Diagrammatic presentation of a relay-cropping system with a long season cereal (Guinea zone dwarf sorghum) (o) and two early maturing crops, a short season cereal (Pennisetum millet or maize) (x) and a late planted legume (o) tried in Nigeria (after Andrews, 1973).

The maize and millet are early maturing, using the first 80 to 90 days of the season, whereas the sorghum, although planted with or just after the early cereals, has a long period of vegetative growth, and floral initiation does not occur until after the early cereals are harvested. Cowpeas are planted mid season in the space vacated by the millet or maize and they mature at the same time, or slightly later, than the sorghum, i.e. at the end of the (moisture) season.

Succesful relay cropping of rice and sweet potato, rice and maize, rice and tobacco, rice and sugarcane and rice and jute are common in Taiwan. The Indian Agricultural Research Institute (1972) has reported on several relay cropping systems, of which combinations such as mung-maize-toria-wheat and mung-maize-potato-wheat were the most important. Relay cropping systems using cassava which are common all over the tropical world because of high yield stability, have been described by Hart (1975) and Beets (1976).

MULTI-STOREY CROPPING

Multi-storey cropping with permanent and annual crops

Trees in coconut, rubber and oil palm plantations are generally quite widely spaced and the trunks only occupy a

20

small fraction of the land surface. Since the tree canopies generally let through most of the light,it is possible to grow crops underneath them. Subject to soil fertility, a second or even third crop can be supported. In an example of three storey cropping, the layer immediately above the ground is occupied by crops such as groundnuts or sweet potatoes; the leaves of papaya or a musaceae crop occupy a level of 2 to 5 meters above the ground and a coconut canopy forms the top at a level of 5 to 15 meters. Nelliat et al (1974) described another interesting multi-storeyed crop combination consisting of coconut and black pepper and coffee and pineapple which is illustrated in Figure 2.7. In Malaysia,intercropping in the two main plan-

Figure 2.7   Multi-storey crop combination of coconut, black pepper, cacao and pineapple at an experimental farm in India. (After Nelliat, et al, 1974).

tation crops rubber and oil palm, is done successfully with maize, upland rice, soya beans, groundnuts and cassava (Benclove, 1975). Associations of rubber and maize or cassava can also be found in Latin America (Morales et al, 1949). The chemical soil fertility must be adequately maintained in these systems since tree crops generally withdraw great quantities of nutrients and annual crops will not produce unless they are given fairly large fertilizer dressings. In contrast to the above, Vidal (1965) described a system in Senegal with the tree Acacia albida. This tree is not planted by the African subsistence farmers but is left standing when the bush is cleared for crop production. The Acacia population may reach forty to fifty trees per hectare and underneath the trees a circle of millet is planted. The millet closest to the tree usually gives a higher yield than that planted further away because nitrogen levels are higher there and also because the tree favourably changes the micro-climate for the millet.

Multi-storey cropping with permanent crops only

Multi-storey cropping of rubber and cacao is reported by Hacquart (1944) in the Congo, by Allen (1955) in Malaysia and by Hunter (1961) in Costa Rica. Associations of rubber and coffee can be found in Malaysia (Allen, 1955), Indonesia (Cramer, 1957), and Costa Rica (Hunter, 1961). Generally, shade requirements for cacao and coffee tend to increase as growth conditions become less favourable and vice versa. These systems are most suitable to areas of low fertility and less favourable conditions. One component of the association can, however, favourably change the micro-climate for the other, and this may improve the growing conditions for both crops. When evaluating the merits of these systems factors that have to be considered include total yields, aggregate income of all crops, access to crops and, ease of management, especially weed control.

A less common system can be found in arid areas where many layers of date palm, apricot and vegetables are traditionally grown in desert oasis (Baldy, 1963). In such plant communities, shading and windbreak effects create a favourable micro-climate for the storeys below; the crop chosen for each successive lower storey should be more mesophytic and less light demanding than the one above (see also Chapter VI).

STRIP OR LANE CROPPING

Strip cropping is defined as growing two or more crops in alternating strips or blocks on the same piece of land at the same time. The difference between this system and intercropping is in the degree to which the two crops interfere with each other. In mixed cropping, species are intimately mixed while in strip cropping, only the plants on the edge of a strip affect the other. The alternating strips should be wide enough to facilitate the use of machinery. The width of the strips also depends on the competitive yield advantage of the one crop and yield disadvantage of the other crop. The major advantage of this system is that the border rows of the tall crop yield 20 to 40 per cent more than rows within the field (Beets, 1976; Pendleton et al, 1963) and that lodging of both crops may be reduced (Beets, 1976). The shorter crop is, however, normally at a disadvantage and its yield is frequently reduced by 5 to 20 per cent (Lang, 1949; Pendleton, 1963). Figure 2.8 shows a strip cropping

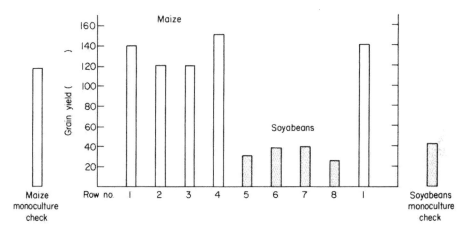

Figure 2.8   Diagrammatic presentation of yields of maize and soya beans obtained in a strip cropping system compared to monoculture checks for both crops (After Pendleton, et al, 1963).

system tried out in the cornbelt of the U.S.A. (Pendleton et al, 1963), in which all maize rows had a higher yield than the monoculture check, while all soya bean rows had lower yields.

Sometimes crops are planted in strips to combat erosion. On long slopes subject to sheet erosion the field may be laid out in narrow strips across the incline, alternating erosion-sensitive and erosion-resistant crops. In such cases, the width of the strips depends primarily upon the degree of slope.

Skip-row systems or skip-furrow planting

The skip-row system, the main advantage of which is that

23

irrigation water can be saved, is practised in low rainfall
areas. According to Sivanappan et al (1976), it is
possible to save 50 per cent of the irrigation water
without significantly reducing yields when such planting
systems are used. Figure 2.9 shows a skip-furrow

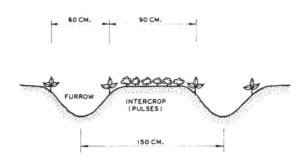

Figure 2.9   Schematic presentation of a cotton-pulses skip-
furrow planting system. (After Sivanappan,
et al, 1976).

irrigation system tested in India (Sivanappan et al, 1976).
In this system the furrows were spaced at 150 cm as opposed
to conventional spacing of 75 cm and a row of cotton was
planted on each side of the furrow leaving a space of 90 cm
between the rows of cotton on the side of the furrow for a
short-term intercrop like pulses. The plant population of
the main crop (cotton) was as high as in a conventional
system and the different planting pattern did not result in
any significant reduction in yield while an additional crop
of pulses is obtained. Similar systems are suitable for
crops such as cotton, sugar cane, castor beans (main crop)
and soya beans, grain sorghum and pulses (intercrop).

Alternating bed system

In areas with periodically waterlogged soils and insufficient water for flooded rice, land can be more intensively
used if the field is prepared in an alternation of low and
high beds. The low beds are used for rice and the high
beds for upland crops. The width and length of the beds
depends on the topography of the field. The major advantage of this method is that the upland crop can be planted
before the harvest of the paddy crop at the end of the long
rains and does not, therefore, suffer from water-logging.
This system is used in Central Java, Indonesia (see Figure
2.10), Vietnam and Taiwan (Hao, 1972).

24

Figure 2.10 The "sorjan" system of Central Java. Upland crops are grown on the raised beds in a mixed cropping arrangement and lowland rice is grown between the beds.

ANNUAL WINDBREAKS

Permanent windbreaks of shrubs and trees are common in France, the U.S.S.R. and the U.S.A. Temporary windbreaks of tall annual crops are less usual but can be advantageous in areas where droughts are common and dry soil and hot winds reduce yields. Windbreaks modify the micro-climate, mainly on the lee side of the windbreak. Figure 2.11 illustrates an association of a tall annual plant (maize) which act as a windbreak and a short annual plant (soya beans) which profits from the change in micro-climate induced by the maize. The mean horizontal wind speed over the soya beans is reduced by the maize barrier.

Annual windbreaks also reduce evaporation from the soil and transpiration from sheltered plants, particularly during hot, dry periods. Consequently, the short plants use less water and are less likely to wilt and the sheltered plants grow taller and produce higher yields.Although there may be no difference in actual amounts of water used, water-use efficiency increases because of the higher yields (Radke and Burrows, 1970). While windbreaks provide good results in arid climates they can also increase yields in semi-arid and even humid areas (Rosenberg, 1975). Almost any combination of crops or varieties can be used provided one is tall and the other is short. The most advantageous combination will depend on the compatibility of cultural practices of the two crops. Tall crops which have been used

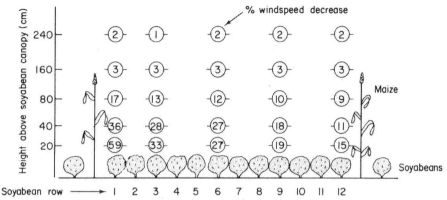

Figure 2.11 Schematic presentation of windbreak cropping system with maize and soya beans. Percentage decrease in the mean horizontal windspread above the soya beans is indicated for five levels above the soya bean canopy (From Radke and Burrows, 1970).

include maize, sugarcane and sunflowers and short crops have included soya beans, sorghum, groundnuts and high-value horticultural crops.

SEQUENTIAL CROPPING SYSTEMS

Sequential cropping is defined as growing more than one crop on the same piece of land with each crop during a different time of the year. Some examples of this are rotations of wheat and soya beans in Taiwan, the U.S.A., India and Zimbabwe. In subtropical areas, soya beans are grown in summer and wheat is grown during the cooler months.

Triple cropping with high yielding varieties of rice attaining total yields of over 24 tons rice per hectare per year is sometimes practised in Southeast Asia. The total growing period required to obtain these total yields is around 340 days with 25 days available for preparing the land. These practices provide some indication of the current upper limit of rice production per unit of land.

Sequential cropping can only be practised in the tropics or subtropics where temperatures are suitable for plant production throughout the year. Other important points of consideration with sequential cropping systems are availability of water and time for land preparation. There are very few areas in the world with sufficient precipitation to support crop growth during each month of the year and irrigation availability is, therefore, often the major constraint to widespread implementation of these systems.

III History of multiple cropping

INTRODUCTION

In early history, before widespread urbanization, the
world's cropping systems were hot as varied as at present.
Most farming was largely on a subsistence basis and there
were less pronounced differences between systems as is the
case with high and low levels of technology. The oxdrawn
plough, hoe, digging stick, sickle, harrow, axe and machete
were as commonly used in Europe as in Africa. Cattle
manure and the growth of legumes to maintain soil fertility
was common, and irrigation, water-lifting systems and a
variety of other fundamental techniques where known in many
parts of the world. Farmers were slow to respond to
technical, economic and environmental changes. The
situation described above still prevails in most tropical
areas - farmers use traditional tools and respond slowly to
change.

EARLY EVIDENCE OF MULTIPLE CROPPING

Since mixed cropping was well suited to this situation, it
was common in most traditional farming systems. In Britain,
mixed cropping of barley and clover was quite common and in
India, the practice of growing mixtures of legumes and
non-legumes was widespread. In 1887, Wallace studied mixed
cropping in India and he found it very advantageous. He
noted the following:

(i)   roots of different species take up different
       nutrients and do therefore not compete with
       each other; and

(ii)   in mixtures of grain and pulse crops, the
       grain crop benefits from the nitrogen secre-
       ted by the pulse.

American Indians practiced mixed cropping of maize and
beans in the eighteen hundreds (Hariot, 1888). Willis
(1914) observed mixed cropping of perennial as well as
annual crops in Ceylon, Malaya and the West Indies and
concluded that mixed cropping practices could well be
advantageous in traditional cropping systems.

Sequential cropping was less common, most probably because irrigation is often required for this cropping system. Gompertz (1927) indicates that in "very early times" there was a form of perennial irrigation at Memphis and Abydos which produced more than one crop a year. The double cropping system practiced involved wheat with a growing period of three months.

Double cropping has long existed in China. In the north, the principal crop was winter wheat, while in the south, it was rice. The development of an early rice variety in the year 1012 triggered a revolution in growing practices and made cultivation of a second crop possible. By the Ming period (1368-1644) cold tolerant varieties were developed which could be planted in mid-summer after spring crops or early rice. As a result of the introduction of these varieties, Kwangtung Province and the southern portion of Fukien Province were reportedly famous for rice. Further north, in Hunan, it was not until the seventeenth and eighteenth centuries that efforts were made to promote second crops (Perkens, 1969).

RECENT HISTORY OF MULTIPLE CROPPING

Although multiple cropping is common in South America and Africa, the recent history of the practice is not well documented for these continents. There has been sporadic interest in the subject and only a few researchers have studied it and there have been no large research programmes designed to study or promote multiple cropping practices. In Asia, the situation is very different. Several countries in Asia have a most interesting history of multiple cropping. Taiwan can serve as an example.

The history of Taiwan's agricultural development is well documented and multiple cropping systems have been a distinctive feature of its development. Multiple cropping was originally brought to the island with the mass migration from the south-eastern part of mainland China which started during the seventeenth century. The migrants not only introduced such crops as rice, sugarcane and sweet potatoes but also water buffaloes, farm tools, new cultural methods and multiple cropping practices which had been applied on their native land for a long period of time. Multiple cropping developed in Taiwan for a variety of reasons, population pressure being the most important. The systems were made possible because of the favourable climate which enables crops to be grown throughout the year. Although the climate is favourable and rainfall reasonably high and distributed, irrigation works have helped the development of multiple cropping. Another important factor was the development of an extremely good early maturing variety of rice ("Taichung" no. 150) in 1938. This variety is excellent for relay-intercropping systems - the most important form of multiple cropping on the island. Special varieties of sugarcane, tobacco, wheat and other crops were also bred over the years. The multiple cropping index for Taiwan (not including green manure) reached 189 - 190 in the years 1965-66, and gradually fell to 175 in 1972. The fall is partly due to the decrease in the area of paddy fields and partly to an increasing shortage of farm labour. Since the beginning of this decade Taiwan's industrialization has lightened the pressure on the land and the farms are less characterized as subsistence enterprises. As the result of these economic developments it is possible that agricultural systems will now develop along Western lines and mixed and relay cropping will become less common.

India is another country where multiple cropping has played an important role in agricultural development. In this heavily populated country, multiple cropping has enjoyed attention since the nineteen thirties. Population was an important factor here' also and the practice was especially common in the Ganges Valley and the Ganges Plain. According to Ganguli (1930) double cropping was not very productive at the time since no fertilizers were available and low soil fertility could not support double cropping. An exception was the deltaic portion of the Ganges Plain where the annual rise of the river left a fertilizing deposit of silt. Under conditions of low soil fertility there is more scope for mixed and relay cropping and in India there has been interest in all-pulse intercropping systems for the areas with limited water supply and low soil fertility since the nineteen forties.

RECENT HISTORY AND DEVELOPMENT OF RESEARCH

Multiple cropping has been a matter of increasing interest in many developing countries during the last two or three decades. Research in this field increased after the introduction of the so-called "green revolution". When the hopes inspired by the green revolution were not met, agriculturalists and policy makers began searching for other means to improve agriculture in the developing world.

Multiple cropping systems research has been done by the large international research institutes such as the Centro Internacional de Agricultura Topical (CIAT) in Columbia, The International Crops Research Institute for the Semi-Arid Tropics (ICRISAT) in India and the International Rice Research Institute (IRRI) in the Philippines.

At IRRI, a multiple cropping research programme was initiated in the 1960s. The main objective of the programme was to develop improved and intensified cropping patterns to increase the welfare of rice farmers in Southeast Asia. Cropping systems technology was organized to use farmers' resources more efficiently in meeting this goal (IRRI Annual Report 1973). Most of the work of IRRI is centred on rice but upland crops such as maize, mung and soya beans are also included in the programme.

Much work has been done on intercropping but relay cropping systems have also been examined. An important aspect of the programme is that advanced, scientific research is done along more practically oriented work. Research is often carried out at different levels of technology; for example, power sources such as hand labour, small tractors and carabao are compared in trials.

The interest in multiple cropping research at CIAT, Columbia is fairly recent. Multiple cropping is, however, widespread in Latin America and it can be expected that the research effort in this part of the world will increase. The study of mixed cropping is an important element of the research undertaken at CIAT. Frances and Flor (1975) have been working on maize and bean varieties, particularly their germplasm, and their usefulness for intercropping systems and have studied cropping systems by variety crop interaction. Their agronomic work has included relative dates of planting and plant population.

At ICRISAT in India multiple cropping research is also fairly recent and has focussed on mixed cropping. The study of crop interrelationships, especially the effects mixtures have on nutrient and moisture uptake, are important elements of the programme. The Institute has more recently started to study whole farming systems.

IV Economic and social aspects

INTRODUCTION

The conditions under which a tropical farmer operates his farming enterprise differ from those in the "western" world. In the tropics, almost all farms are small and subsistence is usually more important to the farmer than cash cropping. The farm operation is based primarily on manual and animal labour. A considerable proportion of the farm output is consumed by the family and the rest of the produce is sold or bartered at nearby markets. This means that a tropical farmer not only measures the "output" of his farm in monetary terms but also in such terms as "foodvalue" and "return per unit of labour".

Under the conditions described above raising production through expanding multiple cropping can only work when a systems or integrated approach is used and when several constraints are removed simultaneously. In this chapter, a number of problems are described in isolation and it should be realized that in few, if any, instances will removing of only one or two constraints result in significant development and expansion of multiple cropping.

Broadly speaking, to increase the productivity of traditional tropical farming systems, two main changes can be made:

(i) raising the level of technology and increasing the level of external inputs; and
(ii) improving marketing and distribution.

The availability of external inputs varies greatly from location to location and directly influences the character of the local farming system. Consequently, the availability of inputs in a certain area must be assessed before the operation of an agricultural system can be understood, changed and improvements recommended. As the level of technology rises and more inputs become available, marketing and distribution usually require improvement. This generally means better storage and transport facilities.

LEVEL OF TECHNOLOGY AND RESOURCES

The inputs used in a farming system can be divided into the

following four groups:

    (i)   natural resources (climate, soil, etc.);
   (ii)   human resources (labour, entrepreneurship, etc.);
 (iii)   external inputs (fertilizer, insecticides, etc.); and
  (iv)   financial resources (credit).

Highest productivity can expected to be attained in areas with fertile soils, high temperatures throughout the year, a high and well distributed rainfall and farmers who have sufficient trained labour, access to external inputs (e.g., fertilizers, high yielding varieties seeds, machinery) and easy access to markets and credit.

If human and financial resources are abundant and the level of technology is high these factors can sometimes compensate for sub-optimal natural resources. For example, the environment can be improved by the introduction of irrigation systems, drainage works and land levelling. When the level of technology is low, however, farmers depend entirely on the existing natural resources. The latter situation is the most common in the tropics where 80 per cent of farmers depend for their survival solely on their own labour (with or without animal power) and the natural soil fertility and rainfall.

It is sometimes possible to increase the productivity of traditional farming systems without introducing external inputs by making better use of available resources. This can, for example, be done by planting at the right time, better weeding and correct plant populations. Unfortunately, productivity increases resulting from these measures are usually small. Therefore, it is normally necessary to introduce some new technology and external inputs.

When introducing new technology and inputs into a traditional farming system the existing system must change in order to accommodate the inputs. This change can either be dramatic or gradual through careful preservation of useful elements of the traditional system and adapting the system to increased quantities of inputs. Generally, when a dramatic change is attempted whereby the entire system is replaced with a new one designed for a high level of external inputs, the change will often not be as complete as was intended and many undesirable elements of the traditional system will remain. The resulting unintentional mixture of traditional and new system often results in an ill-adapted system. On the other hand, when a careful selection is made of the positive elements of the traditional system and these are appropriately combined with external inputs, the result may be a well-adapted system. As illustrated in Figure 4.1, it is in the latter situation that the highest productivity can be expected.

Because of numerous constraints it is often not possible to achieve high levels of technology. It is then necessary

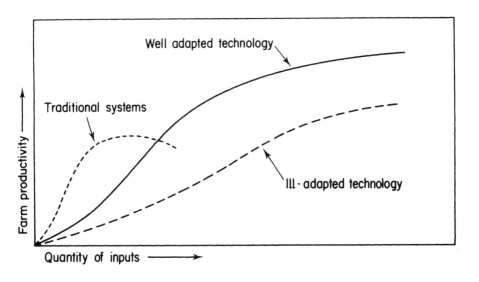

Figure 4.1   Comparative efficiencies of different farming systems with respect to use of inputs. (After Harwood and Price, 1975).

to raise productivity by introducing some basic inputs. This is often a good strategy: when the quantity of inputs into an efficient traditional farming system is increased slightly, farm productivity normally increases sharply (see Figure 4.1). As the quantity of inputs continues to increase, however, the rate of improvement in traditional systems diminishes. This can be illustrated by considering the application of fertilizers. When small quantities of fertilizers are applied to crops grown in a traditional farming system the yields increase. Most traditional varieties are not, however, highly fertilizer responsive and increased applications do not lead to similar increases in yields. Although traditional farming systems respond greatly to small increases in the application of inputs a yield plateau is reached quickly. On the other hand, although high level technology systems respond to high levels of inputs there is a danger that if the technology is not well adapted, the response to inputs will be disappointing.

Mixed and relay cropping are important elements of traditional cropping systems. It is better to adapt these systems to increased quantities of inputs than to replace them with sole cropping. This can be easily done by changing plant populations and plant configuration; changes which normally mean that a mixed crop responds to the application of fertilizers. Mixed cropping systems often use more soil moisture than sole crops and they, therefore, are responsive to irrigation. Hence, when irrigation is introduced it is not necessary to change from mixed to sole cropping.

Sequential cropping systems need a high level of technology and can often not be practised without fairly high levels of external inputs, particularly fertilizers. Also, irrigation is often a prerequisite for these systems.

MANAGERIAL ABILITY OF THE FARMER AND TRADITIONS

The level of education and the farmers' understanding of their environment and how best to exploit it greatly influences the character of the local cropping system. In many parts of Southeast Asia farmers seem (or seemed) to live in harmony with their environment. The irrigated rice cultivation on terraces in Bali and parts of the Philippines are examples of man "mastering" his environment. On the other hand, vast, dry barren and eroded lands in many parts of Africa show how greatly people can abuse and misunderstand their environment. While an equilibrium between the biological and cultural environments was usually found in the past, this equilibrium was often lost due to over population and other stresses.

Traditions and certain superstitions play an important role in farming practices in the tropics. A good example is the ownership of cattle in Africa. In many regions of Africa cattle are not kept for economic reasons. Rather than being used to provide milk, meat and hides, cattle are used as a symbol of wealth and status. The animals are kept mainly to buy wives and to pay for dowries. In such circumstances their usefulness in terms of production is of minor or no importance.

Cruz and Alviar (1975) write that in Quezon Province in the Philippines planting days are based on the "Honorio Lopez Calendar" which suggests that plants produce low yields when planted on days of the first or last moon quarter. Hence, the farmers in the region only plant during full moons. "Time of planting" is, however, often crucial in multiple cropping and a delay in planting crops in relay or sequential cropping systems by only a few days can already reduce yields.

The cropping calendar and the calendar of social events are often closely interlinked. Multiple cropping may cause a significant change in the accustomed rhythm of life since relay and sequential cropping often occur in what used to be the off-season. This period between harvesting and planting crops is often the time traditionally used for celebrating marriages, visiting relatives, and so on. When intensive multiple cropping systems are introduced, extra crops should normally be raised during these periods which interferes with local customs and may lead to social tensions. According to Singh and Kumar (1972) these considerations are important in India. De Sapir (1970) mentions the same problem in Africa.

Dietary preferences can sometimes determine the character

34

of a cropping system. In many parts of the world, and especially in Asia, rice is preferred above all other staple food crops and farmers are, therefore, reluctant to grow anything but paddy.

A change in cropping patterns often causes a change in social customs which may or may not be acceptable. When the farmer clearly sees the advantages of a new system, when he has tangible wants, or when he has simply run out of food, changes will be acceptable. On the other hand, when changes are drastic and greatly affect social customs and his needs are not too pressing he will often say: "what would I do if I worked harder and earned more money?" In such cases the new system may not be adopted.

When the farmer is progressive and determined to improve his condition, his success depends on his ability to introduce improved cropping systems. If his level of education and general understanding is high enough to comprehend how new, higher yielding systems can be introduced, and when he is willing to provide the necessary labour, his success depends on the supply of inputs. Whether inputs are available to him depends, in turn, on the infrastructure of the region and prevailing economic factors.

INFRASTRUCTURE

Introduction

The lack of physical infrastructure (e.g.,roads, irrigation works, buildings) and weak agricultural institutions (e.g., farmers' organizations, credit unions), often explain why agricultural development is slow. Generally, only farmers living in a market economy can expect to have access to the farm inputs necessary to increase output. In order to develop a region and make it more productive it is necessary that the infrastructural development be balanced. For example, neither roads nor irrigation works on their own can increase the production potential of an area. Both are necessary.

Roads

Roads can spearhead development and often the greatest single factor facilitating the integration of an area with a market economy is the development of the transportation system. Highway systems and well designed networks of feeder roads not only facilitate transportation but also reduce the maintenance costs of motor vehicles and extend the life of cars and trucks. Unfortunately, when new transportation systems are planned there is often too little coordination with agricultural planners to ensure that a total package of programmes necessary for the complementary agricultural development is provided.

Irrigation

In most parts of the world year-round cropping cannot be sustained under rainfed conditions; irrigation is therefore often a prerequisite for sequential cropping. In many areas the extent of sequential cropping is positively correlated with irrigation. In Taiwan, for example, the Multiple Cropping Index for irrigated rice is 225 and for rainfed rice 140. (Wang and Yu, 1975). East Asia, with the largest percentage of arable land under irrigation in Asia, has cropping indices of between 150 and 200 (Chao, 1975). The irrigated areas in Indonesia (Oshima, 1973), Thailand (Manu, 1975) and India (Rao, 1975) have cropping indices of between 125 and 150. By contrast the Philippines, which has little irrigation on a countrywide basis,has a cropping index of approximately 100 (Harwood and Price, 1976).

Irrigated agriculture has a long history in Asia and many farmers are familiar with it. Consequently, sequential cropping systems are often adopted relatively quickly after irrigation is introduced. Because output under sequential cropping is far greater than under mono-cropping,irrigation is often economically feasible in Asia. In fact, the introduction of irrigation may be the only way to increase productivity.

Irrigation is relatively new in most of Africa and farmers have little experience with irrigated cropping systems. Although irrigation opens up possibilities for new crops and increased productivity, it is often doubtful whether farmers will be able to adopt cropping patterns which make use of the water. Further, in Africa there is still considerable scope for increasing agricultural productivity by opening up new land and increasing the yield of present rainfed agriculture. Irrigation has, therefore, often proved to be only marginally economical in Africa.

In Latin America, rainfall can generally sustain at least two crops a year. Land is usually not scarce and irrigation is only necessary and feasible in small areas on this continent.

Market and distribution

The average farm size in the tropics is small. Small farms are associated with high marketing costs for agricultural products since overheads are proportionately higher. Individual farmers cannot afford to transport their own produce and are dependent on numerous middlemen to do the transportation and distribution of the produce. These middlemen are often poorly organized and take excessive commissions.

When there is no network of assembly and collecting stations and when no processing plants such as rice mills, marketing stalls and centres exist, little economic growth can be expected.

There is often a lack of storage facilities in the tropics. At present, according to FAO estimates, developing countries lose 30 percent of their potential food production to pests. A great share is lost during storage. When crop produce cannot be stored, farmers have to sell at low prices immediately after harvest which often makes it difficult for these farmers to enter the market economy. Consequently, they remain producing at a subsistence level where little progress can be expected. Under conditions where a farmer cannot rise his status as a subsistence farmer, mixed cropping systems are often favoured. Indeed, under these circumstances these systems with their lower "risk factor" seem appropriate. Multiple cropping practices are also advantageous because of the greater spread of the harvest over the year which reduces storage time for the crops.

Capital and credit

It is difficult for small-scale farmers to substantially increase production and farm income without capital. Operating as well as investment capital is normally provided through outside sources, often through low interest government loans.

Throughout the tropics the rural interest rates are high and sometimes local moneylenders set rates which may exceed 100 percent per annum. Price (1973) estimates the average interest rates on loans from all sources in northeast Thailand was 80 percent per annum, and according to Huang (1975) the average farm rental in Taiwan, before the land reform, was fixed at approximately 50 per cent of the total annual main crop yield. Vigo (1965) states that the main source of funds in Northern Nigeria is from local moneylenders who charge rates of interest ranging from 50 to 100 per cent. This situation is also common elsewhere in Africa.

Major schemes such as construction of irrigation projects and major roadworks can generally only be financed with the aid of regional development banks or the national government. This means that rural development programmes are often dependent on outside sources of capital. On the other hand, development can also be initiated by local capital. Examples of projects which can be financed by local sources include storage facilities, introduction of better seeds and fertilizers and construction of minor access roads.

Institutional factors

A good community-level structure and a fundamental nationwide institutional establishment are equally important and crucial to the agricultural development of a region. It is almost impossible for a rural community to manipulate its own complicated production system without assistance from an outside community. Several nationwide institutions are necessary for agricultural development. The development of research and extension services are of obvious importance, and an institution that can provide credit is generally a prerequisite for development.

Presently, land ownership is thought to be an important factor in development and land reform is a high contention factor in the development discussions in many countries. There are some examples of situations where land reform has worked but it is by no means clear whether and how it should be done and what effects it has on production. Nevertheless, it can be argued that land ownership is crucial in creating an environment necessary for promoting agricultural development because uneven land ownership may act as a disincentive for production. On the other hand, more equitable land distribution may result in a progressive and achievement oriented rural society in which there is a need for multiple cropping.

In Africa, land ownership structures sometimes hinder agricultural development because land belongs to the whole community and no single individual is responsible for its use which may result in limited production incentives.

POPULATION AND FARM TYPE

Introduction

In principle, there is a maximum population for any given area which could be supported indefinitely under a traditional cropping system. Without outside influence, the physical environment and the level of technology would determine the population which could be supported in the area. During the past century, outside influences have contributed to marked population increases all over the tropical world. Production systems therefore have to be changed so that the limited natural resources available can support the increased population. The effects of population pressure include decreasing farm size, increasing labour supply and a greater demand for food. Farmers, in turn, must increase their cropping intensities and production to the point where maximum use is made of the resources available. Multiple cropping systems with high food outputs and high labour demands are, therefore, widespread in areas with population pressure. Often, there is a direct correlation between the expansion of multiple cropping and population. Examples of such correlations are given by Revelle and Thomas (1970) for Bangladesh and by Herrera and Harwood (1973) for Asia generally.

There are three basic factors to consider when analyzing labour in rural areas in the tropics:

(i) the need for employment of landless labourers throughout the year;

(ii) the need for additional labour at key peak periods such as weeding and harvesting; and

(iii) the need to reduce the seasonality of employment and to spread the workload more evenly during the year.

In most areas the greater part of farm labour is provided by the family. According to IRRI (1971) family labour made up more than 90 percent of the work force in the Philippines at Calen, Batangas, and according to Norman (1973) almost all the work in farms in Northern Nigeria is done by family labour. During peak periods casual labour can be hired or exchanged between families, something which is quite common in parts of Africa.

As shown in Table 4.1 labour requirements vary considerably for different multiple cropping systems and in planning cropping patterns it is important to assess both total and peak labour requirements:

Table 4.1
Labour Requirements of Different Cropping Patterns in Taiwan

Cropping Pattern	Labour Requirements (man-day/ha)
Rice-rice	192.5
Rice-rice-maize	304.3
Rice-rice-sweet potato	312.0
Rice-rice-soya beans	325.1
Rice-rice-flax	239.3
Rice-rice-vegetables	422.6
Rice-rice-tobacco	968.0
Rice-sweet potato	217.1

Source: Research Institute of Agricultural Economics, National Chungsing University, 1972

Birowo (1975) found that in West Java, Indonesia, a cropping pattern involving rice, groundnuts and soya beans consistently provided the largest degree of employment and the highest level of labour income both for family labour and aggregate family and hired labour. The labour requirements for five multiple cropping systems in West Java are presented in Table 4.2.

Table 4.2
Use of Labour For One Hectare in Five Cropping Systems in Two
Sample Villages in West Java, Indonesia

Item	Cropping Systems				
	Rice-Rice	Rice-Groundnuts	Rice-Maize	Rice-Soya beans	Rice-Mung beans
Annual Labour Requirement					
a. Man-days	490	423	680	705	678
b. % of available supply a/	60	52	84	87	83
Monthly Use of Labour, % of Available Supply					
October	25	86	75	59	128
November	28	47	46	12	22
December	84	86	51	51	81
January	46	48	137	137	137
February	23	223	65	66	66
March	4	1	24	24	24
April	105	59	25	25	25
May	129	115	75	96	96
June	90	56	176	155	152
July	53	89	127	157	107
August	78	8	118	211	105
September	37	4	81	53	53

a/ There are on average 2.7 family labourers per farm whose
work is equivalent to 810 man-days per year or 67.5
man-days per month.

Source: Birowo, 1975.

The Indian Agricultural Research Institute (1972) analyzed
the relative employment potential of different crop sequences and found that by raising the intensity of cropping
from two crops to three crops a year, the employment potential was raised by 40-50 per cent, and when cropping intensity was further increased and a quadruple system of mung,
maize, potato and wheat was used, there was an increase of
80-140 per cent in the employment potential over a double
cropping system.

Increased food production and employment creation are
official national objectives in most tropical countries.
These examples suggest that intensive cropping systems have
the potential to make important contributions toward the

achievement of both objectives.

Since crop production is seasonal in nature, labour requirements are also seasonal. The actual distribution depends
on the cropping pattern. Representative labour profiles
for mono and mixed cultures are shown in Figure 4.2.

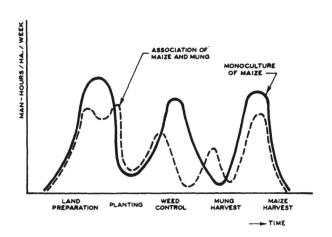

Figure 4.2 Labour profiles for a maize monoculture versus a
mixed culture of maize and mung beans in a hand
operated cropping system in the Philippines
(IRRI, 1975).

Although mixed cropping has generally fewer labour peaks
than monocropping, the period for land preparation is an
exception. It is, however, possible to stagger planting to
some extent, and in that case, and in pure forms of relay
cropping, labour peaks for mixed cropping are lower than
for single crops. Peak labour requirements for weed
control and harvesting are low for crop associations since
the crop components do not have to be weeded or harvested
at the same time. As a rule, the more crops are planted,
the more likely the workload is spread evenly over the
year. Although peak labour requirements for mixed cropping
systems are generally smaller than for monoculture systems,
the total labour requirements per hectare is about 60 per
cent higher. Because peak labour requirements are usually
a constraint on output in tropical farming systems, mixed
cropping is preferred.

Sequential cropping systems normally have pronounced
labour peaks. The period which is specially critical is
the turn-around period (the time between harvesting and
planting the subsequent crop). In Southeast Asia, the

41

adoption of double cropping of rice is often hindered by farmers inability to have short turn-around periods. The reason usually given is shortage of labour. A short turn-around period is necessary since the two crops of rice can only be grown during the period when water is available, usually 6 to 7 months which coincides with the growing period of two rice crops. Therefore, little time is left for land preparation between the two crops. While this problem can be somewhat alleviated by staggered planting, solving it completely can often only be done by introducing mechanization.

In summary the profile of labour availability has strong implications for the design of alternative cropping patterns. Multiple cropping patterns can often be designed in such a way that labour requirements are more evenly distributed than those of monocultures.

Labour and tools

When considering the productivity of labour it is important to know what type of tools are available. The output of a man planting with a stick will be lower than the output of the same man planting with a small hand planter. Similarly, the output of a man ploughing with a small hand tractor will be higher than the same man ploughing with the aid of a draft animal. Land preparation done solely by hand labour takes a long time, and frequently over two months for an average field. On the other hand, with the aid of draft animals or tractors, the preparatory time, is much shorter.

When weather conditions are not favourable for field operations all of the time, hand labour will take comparatively longer since the work can only be carried out during the periods when weather conditions are favourable. Because the length of the preparatory time is critical, the introduction of a double cropping system in areas where formerly a single crop was planted may fail if sufficient labour is not available during the period available for land preparation.

In countries with significant unemployment and underemployment, multiple cropping offers great potential for increased food production. As labour becomes scarce, less labour intensive systems will have to be found, mechanization will have to be introduced and, perhaps, the potential for mixed cropping will be reduced. On the other hand, mechanization is likely to increase the feasibility of sequential cropping.

FARM SIZE

Small farms are very common in the tropical world - one third of farms in Southeast Asia are less than 0.5 hectare in size, one half of all farms are less than 1 hectare and three-quarters are less than 2 hectares. The average farm size in Southeast Asia is 1.8 hectares, compared to 1.1 hectares in East Asia and 2.4 in South Asia (Harwood and Price, 1976). In Latin America, the farms are somewhat larger, but still very small compared to Europe and North America (120 hectares). According to Pinchinat, et al (1976) a small farmer in South America is defined as one who operates a production unit of less than 7 hectares and practices traditional crop husbandry methods. The majority of small farmers may be classified as the poorest group in rural tropical America.

In Africa,the land area available to farmers is frequently quite large, but the area actually used is small since most of the land is left fallow.

In Taiwan, it has been recognized that small farm size hinders increasing economic efficiency (Wang and Yu, 1975). This does not, however, apply to tropical developing countries. Two factors that play a dominant role in this question are the degree of mechanization and social and political constraints. In order to increase the level of mechanization it is often desirable to increase the field and farm size. Mechanization is, however, still not an important factor in the tropics. In many cases, redistribution of land and increasing farm size are not possible for social and political reasons.

It seems, therefore, unavoidable that advances in crop production in the tropics will have to be made on relatively small farms. This will only be possible if the small areas of land are intensively utilized by multiple cropping.

DEMAND, PRICES AND FARM INCOME

The development of new markets has a great influence on multiple cropping. In Asia, for example, the Japanese market has created a demand for products such as cassava chips, soya beans, maize and sorghum. When market outlets are established, and prices are attractive, farmers will more readily accept a cash crop than when markets are unassured and prices fluctuating. The new cash crop will often fit into a double or relay cropping system.

Multiple cropping generally leads to higher production and therefore to higher farm incomes. Andrews (1971) reports that in Nigeria relay cropping and intercropping gave 59 per cent and 80 per cent more return per acre, respectively, than a sole crop of sorghum, the increase coming mainly from higher cereal yield. Syarifuddin et al (1973) found that in Indonesia a mixed cropping system of maize and groundnuts gave a net return 70 per cent higher than a monoculture maize crop. The Indian Agricultural Research Institute (1972) undertook a Cost/Benefit analysis of three sequential cropping systems and found, as shown in Table 4.3, that costs of production increase with increase in

Table 4.3
Economic Analysis of Three Sequential
Cropping Systems In India

Cropping System	Crop	Cost per ha (Rs)	Gross Income per ha (Rs)	Net Return per ha (Rs)	Net Return per Rupee (Rs)
1	2	3	4	5	6
DOUBLE CROPPING					
	Maize	908	2,972	2,064	
	Wheat	1,281	4,632	3,351	
	Total	2,189	7,604	5,415	2.47
TRIPLE CROPPING					
	Mung	381	1,560	1,179	
	Maize	908	2,972	2,064	
	Wheat	1,281	4,432	3,151	
	Total	2,570	8,964	6,394	2.49
QUADRUPLE CROPPING					
	Mung	381	1,670	1,289	
	Maize	908	2,942	2,034	
	Potato	1,651	6,666	5,015	
	Wheat	1,237	3,860	2,623	
	Total	4,177	14,894	10,961	2.62

Source: Indian Agricultural Research Institute, 1972.

cropping intensity. For example, the cost of production of a double cropping system of maize and wheat was calculated to be Rs. 2,189 per hectare, while for a triple cropping system involving mung, maize and wheat it was Rs. 2,570 per hectare. For a quadruple system the costs were even higher. When, however, net profits per hectare per annum were considered, it was found that quadruple cropping of mung, maize, potato and wheat was the most profitable giving an income as high as Rs. 10,961 per hectare. When the cropping systems were ranked according to costs of production and return on investment, the relative position of the different cropping systems altered. The double cropping system of maize and wheat gave a net profit of Rs. 2.47 per rupee invested, while the triple cropping system gave a profit of Rs. 2.49 per Rupee invested.

CONCLUSIONS

The conditions under which tropical subsistence farmers operate their farming enterprises are distinctly different from those in the "Western" world. The level of technology at which most multiple cropping systems are practised is quite low. Mixed cropping is advantageous at low levels of technology but is less so at higher levels. Sequential cropping systems need relatively high levels of external inputs. Labour surpluses and small farm sizes often make multiple cropping both possible or necessary. When there are no alternative employment possibilities for labour, an appropriate shadow price should be used to reflect the economic value of this input. Inadequate infrastructural support, absence of roads and irrigation works all determine the conditions under which the tropical farmers work. When assessing the possibilities for improvement in crop production in an area, however, the farmer, his level of understanding and his incentive to change are of overriding importance. The potential benefits of multiple cropping practices must always be viewed in the context in which it will be applied. Like other development initiatives, attempts to develop multiple cropping will only succeed if the local socio-economic conditions and constraints are taken fully into account.

V Agro-technical characteristics of multiple cropping systems

INTRODUCTION

The growth and development of a crop, which is an aggregation of individual plants of the <u>same</u> species, can be regarded as a system with the following components:

    (i) the characteristics of the plant species;
    (ii) the functioning of the plant during its development; and
    (iii) the plant environment.

It is the interplay of these components with which agriculturalists are normally concerned.

Since modern crop varieties have a great degree of genetic similarity among individual plants, the characteristics and the functioning of the single plant usually give a good indication of the characteristics of monoculture crops. In such systems the crop-environment interaction is more important than the interaction between individual plants within a crop. On the other hand, in multiple cropping, the interaction between the components of the crop - between plants of different species - is very important. At the same time, crop-environment interaction remains as important as in monocultures, and consequently, understanding the interrelationships between the physiological activities and the environment of a multi-species crop is more difficult than for a monoculture crop.

In sequential cropping, the influences that different crops have on each other is generally not very great. These can be summarized as follows:

    (i)   A crop can affect the soil structure for the following crop in the rotation. An example of a crop that leaves behind a good soil structure is soya beans, and examples of crops which make land preparation and planting of a following crop difficult are cotton and rice;

    (ii)   Crops that are host to soil borne pests and diseases may result in a build-up of pests and pathogens in the soil which may affect the following crop. Tobacco, for example, leaves behind nematodes. On the other hand, marigolds have a positive effect by killing or suppressing nematodes;

(iii) Some crops require large amounts of soil nutrients or soil moisture and as a result, the following crop may suffer shortages. For example, crops such as maize and cassava require large amounts of nutrients, leaving behind a depleted soil.

(iv) Leguminous crops use little or no nitrogen and sometimes even produce nitrogen which can be taken up by a following crop.

In mixed and relay cropping, the above factors also apply. However, since the different species are physically closer and more intimate, there are more factors that play a role. These factors can be summarized as follows:

(i) The stature of the components of the crop association;

(ii) Growing habits and growing speeds; and

(iii) The competitive power of species.

Because of these factors, the selection of crops and varieties is important in all multiple cropping systems.

CROP AND VARIETY SELECTION

Crop varieties used in sequential cropping systems should be quite uniform - the individual plants should all mature during a set period of time. This can be achieved by planting a hybrid variety which is photoperiod insensitive. The latter characteristic not only assures maturity after a set number of days after planting, but it also means that the variety can be planted at any time of the year. Early maturity is another desirable characteristic. It permits a more intensive organization and greater flexibility, especially in selecting planting times.

Sequential cropping of legumes and cereals is a widespread practice and is especially advantageous for soil fertility maintenance. Sequential and continuous cropping of low land rice is quite common in parts of Asia without water constraints. It is avantageous for soil management reasons - it is difficult to cultivate puddled anaerobic lowland soils for upland crops and it is, therefore, difficult to alternate upland crops and lowland rice.

Crop varieties used in mixed cropping systems should have a high plasticity - i.e. they should give fairly stable yields over a wide range of plant populations. This allows flexibility for varying the crop proportions without serious loss in yield. An example is two maize/groundnut associations; one with low and one with high maize populations. If the maize variety has a high plasticity, the yields of the maize will be nearly the same in the two associations because the low plant population is compensated for by higher cob weight. On the other hand, if the maize variety has a low plasticity and only gives yields at a narrow range of populations, it is less suitable for mixed cropping.

Shade tolerance is an important characteristic for short statured plants. Shade tolerance of legume genotypes varies and screening for the most tolerant types is therefore always useful.

The most common mixed cropping associations are those of a legume and a non-legume, often a cereal. The outstanding fact of the legume/non-legume association is that usually neither crop gives as large a yield in mixed cultures as when grown alone, although normally the combined yield is higher than when either is grown as a sole crop. On the whole, compared to pure stands, legume yields are more depressed in an association than are those of cereals. This is illustrated in Table 5.1.

Table 5.1
The Reduction in Yield of Legumes when grown in
Legume/Non-Legume Associations

Type of Association	Reduction (%)	Reference
Maize and groundnuts	20-30	Syarifuddin, et al (1973)
Maize and mung beans	30-34	Syarifuddin, et al (1973)
Maize and soya beans	20-40	Beets (1977)
Castor beans and groundnuts	11-43	Evans and Sreedharan (1961)
Castor beans and soya beans	11-35	Evans and Sreedharan (1961)
Millet and beans	62	Paul and Joachim (1974)
Wheat and lentils	64	Papadadis (1940)
Setaria and pigeon peas	0	Krantz, et al (1976)

The selection of the components of an association is important. Crops and varieties can be screened for their suitability for growth in associations and, of course, crops specially suited to multiple cropping can be selected and bred. The latter has not been done often and there is considerable scope for breeding varieties for specific cropping patterns. This would be in contrast to the present practice of only breeding varieties which give maximum yields under optimum environmental and management levels found in monoculture systems.

PLANT POPULATIONS AND SPATIAL ARRANGEMENTS

Crop yield is a function of yield per plant and number of plants per unit area. In commercial agriculture "the crop" is normally a community of individual suppressed plants (Donald, 1963). Under these conditions yield per plant is relatively low, but since the number of plants per unit area is high, the total yield per unit area may also be high. The number of plants of a certain genotype that can be advantageously planted per unit area depends on the environmental resources. When only limited resources are available, the plant population should be low; when there is an abundance of resources, the optimum population can be high. In mixed cropping, plant populations should be optimum for mixed cropping patterns to be advantageous. As shown in Figure 5.1, the environment should be "saturated",

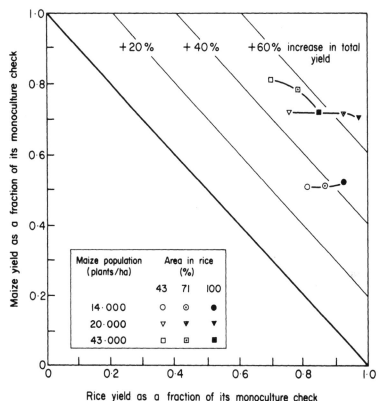

Figure 5.1   Effect of maize population and proportion of the area under rice on intercrop productivity.
(From IRRI, 1973).

in other words, there should be a great degree of "pressure" on the environment. When maize and upland rice

49

were planted in associations, higher total yields were
obtained by associations that had a high total number of
maize plants per hectare (20.000 and 43.000) than with a
low maize plant population (14.000 plants/ha.) Total yield
also increased as the interplanted rice population was increased especially for the treatments with low maize populations, indicating that the environment was not saturated
with the low maize population.

As shown in Figure 5.2, when the number of plants of a

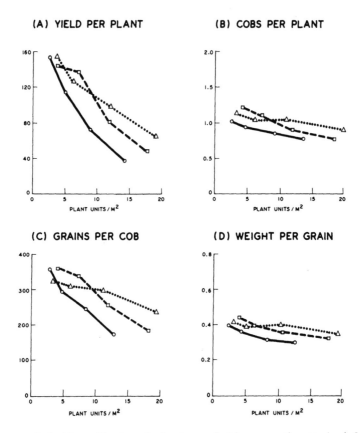

Figure 5.2   The effects of plant populations on the seed yield
components of maize   alone; Mb   2/3 Maize +
1/3 Beans; Mb   1/3 Maize + 2/3 Beans when grown in
a replacement series.
(From Willey and Osiru, 1972).

component of an association is high, the yield per plant is
lower than when there are only relatively few plants. When
maize is grown as a sole crop, the yield per plant, number
of cobs per plant, grains per cob and weight per grain, are
lower than when fewer maize plants are grown in association
with beans. Since the fewer maize plants have more space

available, they grow larger. In most cases, however, the yield per unit area will be lower, since the increased yield per plant does not fully compensate for the decreased number of plants. The yield of the other species in the association will, however, compensate for this yield loss. When the population of one species of an association is reduced, and at the same time the population of the other crop in the association is increased, the contribution to the yield of the association by the first species will decrease and the contribution of the second species will increase. Hence, there is a production shift from the one species to the other. Although there will be no effect on total yield when the environment is saturated; when the number of plants per unit area is too low to exploit fully the resources, the total yield may change when the individual crop populations are changed.

The way a given number of plants is laid out in the field influences the growth, development and yield of the individual plants, as well as the crop as a whole. Equidistant spacing or square planting gives the minimum competitive effect on neighbours since the distance to the neighbours is maximized and this, theoretically, leads to maximum plant yield (Donald, 1963). Since competition is a very important aspect of multiple cropping, planting patterns are also important. Although equidistant spacing is not always possible for management reasons, when crops are planted in rows, many patterns and configurations are possible. Figure 5.3 illustrates examples of desirable arrangements for planting a tall plant and a short plant in association. The crops are planted in rows for management reasons and the spacing of the short crop is equidistant.

The yield of a mixed crop is a function of all factors discussed above and their interactions. The two main factors determining the yield of an association are the proportion of species in the mixture and the populations of the species. It is not possible to compare plant populations of different species directly because different species have different statures which occupy different areas. When optimum monoculture populations are considered, however, it is possible to compare populations of different species by using the concept of "plant unit" which is defined as the number of plants of a certain species that occupy a given land area. Crop associations can best be discussed in terms of plant units and spatial arrangement or plant lay-outs. Willey and Osiru (1972) give a good example of such an approach. They grew different mixtures of maize and beans. The optimum number of bean plants in monocultures was twice the number of maize plants. Thus, when forming the mixtures, one maize plant was regarded as being equivalent to two bean plants. One maize plant and two bean plants are regarded as "one crop unit". When this system is used, the proportion of crops of which the lay-out is given in Figure 5.3.A is one-third maize and two-third beans while in Figure 5.3.B the proportions are reversed.

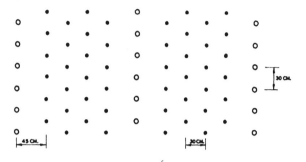

(A) ONE-THIRD TALL SPECIES (E.G. MAIZE) + TWO-THIRDS SHORT SPECIES (E.G. BEANS)

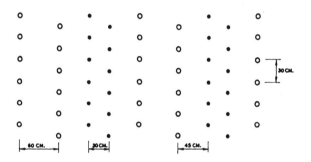

(B) TWO-THIRDS TALL SPECIES + ONE-THIRD SHORT SPECIES

Figure 5.3 Diagrammatic presentation of desirable spatial arrangements of crop rows of an association of a tall plant (o) and a short statured plant (.).

Beets (1976) used a similar approach in studying maize and soya beans in different associations. He postulated that when a maize row bordered a soya bean row, two-third of the space between the respective rows were used by the maize, while one-third of the area was effectively used by the soya beans. Thus, the different proportions were obtained by using a space or area approach. The plant populations, proportions and lay-out of rows are shown in Figure 5.4. Most combinations are mixed cropping systems while others, due to spatial arrangements, are less intimate mixtures and can, therefore, not be considered as mixed cropping systems, but are strip cropping systems.

The proximity of species in mixed cropping systems is important because it affects the degree of intra- and interspecies competition. Interspecies competition (competition between plants of different species) is greater when plants are intimately arranged than when there

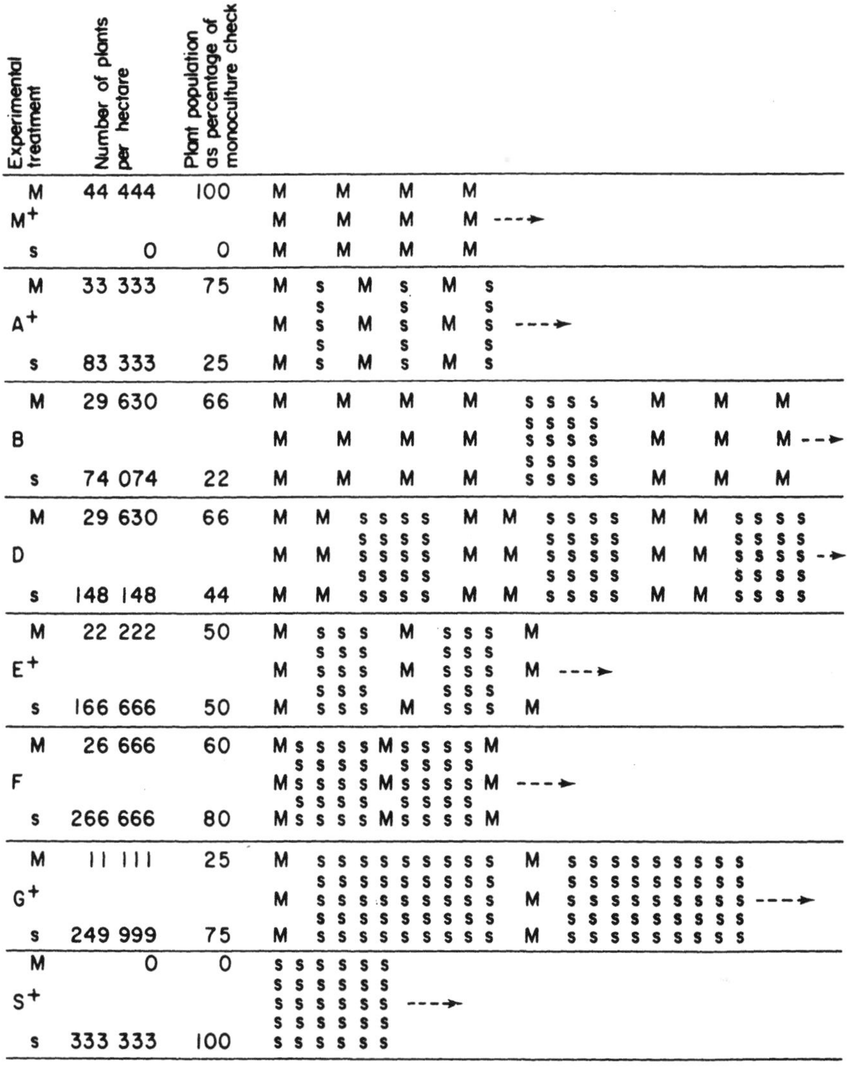

Experimental treatment	Number of plants per hectare	Plant population as percentage of monoculture check	Diagram
M	44 444	100	
M+			
s	0	0	
M	33 333	75	
A+			
s	83 333	25	
M	29 630	66	
B			
s	74 074	22	
M	29 630	66	
D			
s	148 148	44	
M	22 222	50	
E+			
s	166 666	50	
M	26 666	60	
F			
s	266 666	80	
M	11 111	25	
G+			
s	249 999	75	
M	0	0	
S+			
s	333 333	100	

```
M      M     M     M     M
M+           M     M     M     M --->
s            M     M     M     M

M      M s   M s   M s
             s     s     s
A+     M s   M s   M s   --->
             s     s     s
s      M s   M s   M s

M      M     M     M     M   s s s s   M   M   M
                               s s s s
B      M     M     M     M   s s s s   M   M   M --->
                               s s s s
s      M     M     M     M   s s s s   M   M   M

M      M M   s s s s   M M   s s s s   M M   s s s s
                 s s s s         s s s s         s s s s
D      M M   s s s s   M M   s s s s   M M   s s s s -->
                 s s s s         s s s s         s s s s
s      M M   s s s s   M M   s s s s   M M   s s s s

M      M   s s s   M   s s s   M
                 s s s     s s s
E+     M   s s s   M   s s s   M --->
                 s s s     s s s
s      M   s s s   M   s s s   M

M      M s s s s M s s s s M
             s s s s   s s s s
F      M s s s s M s s s s M --->
             s s s s   s s s s
s      M s s s s M s s s s M

M      M   s s s s s s s s s   M   s s s s s s s s s
               s s s s s s s s         s s s s s s s s
G+     M   s s s s s s s s s   M   s s s s s s s s s --->
               s s s s s s s s         s s s s s s s s
s      M   s s s s s s s s s   M   s s s s s s s s s

M      s s s s s s
           s s s s s s
S+     s s s s s s --->
           s s s s s s
s      s s s s s s
```

Note : M = Maize   s = Soyabeans

Figure 5.4 Diagrammatic presentation of treatments of a mixed cropping trial with maize and soya beans. The in-row spacing for maize in all treatments is 25 cm and the in-row spacing for soya beans 10 cm. Row spacings of 90, 60 and 30 cm have been used (the diagrammes are to scale). The + marked treatments are components of a replacement series. (From Beets, 1977).

is less contact between the species. Theoretically, the higher the inter/intra species competition ratio, the more advantageous mixed cropping is, because plants of different species compete less with each other than plants of the same species. In practice, the yield of a mixed cropping system is not only affected by the above but also by such factors as changes in micro-climate and changes in pest and disease occurance induced by the cropping system. All of these factors are influenced by spatial arrangement of rows and individual plants.

Possible differences between obtaining mixtures by alternating rows of species and by mixing species within the row have been reported in numerous publications. Herrera and Harwood (1974), for example, found that yields of rows of maize planted between rice spaced at 1.4 metres were higher than when the rows of maize were spaced 2.8 metres. The row arrangements of this trial are diagrammatically represented in Figure 5.5.A. In the experiment, both maize row-spacing and the number of rice rows between the maize was varied, thus obtaining different areas under rice. In a similar experiment the areas under rice were not varied, but the row arrangement was changed. In this case there was no alteration of a certain number of rice rows between single rows of maize, but two or more maize rows were planted next to each other thus obtaining "units" of maize and "units" of rice. (See Figure 5.5.B.) The least intimate association was obtained by planting three rows of maize alongside three units of rice rows; each rice unit containing 5 rows (3 X 3); and the most intimate mixture was obtained by planting two rows of rice between single rows of maize (1/2 X 1/2). When the row arrangements were changed from 3 X 3 to 1/2 X 1/2, the total productivity increased from 50 per cent to 190 per cent of the monoculture check. Hence, maximum productivity was attained when interspecific competition was highest.

In contrast to the above, it has often been found that in mixtures of maize and legumes, maximum productivity is attained at low maize populations (Evans, 1960; Beets, 1976). This is because the widely spaced rows of maize act as a windbreak for the legume thereby reducing transpiration which leads to higher yields under conditions of slight moisture stress.

An other aspect of plant arrangement which may be of importance is the direction of rows (North-South or East-West). According to Donald (1963), who reviewed the literature on this subject, yields are generally greater with North-South rows than with East-West rows. This is likely due to differences in the light regimes, with superior lighting in North-South rows, as compared with the poor lighting on the North side of East-West rows (for Northern latitudes; South of the Equator the situation will be reversed). Workers at IRRI (1975) and Samson and Harwood (1975) who experimented with different row directions of mixed cultures of maize and rice found no significant differences between treatments. Pendleton, et al (1963) who experimented with strip cropping systems of maize and soya beans, found no significant effect on yield of either maize or soya beans from strip planting. Some interesting

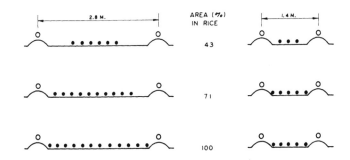

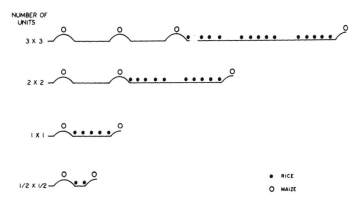

Figure 5.5   Diagrammatic presentation of plant arrangements of
mixed cultures of maize (o) and rice (.). Arrangements in Figure 5A are designed to compare intra-
and inter-species competition and arrangement in
Figure 5B are designed to compare different
arrangements for total productivity. (From Herrera
and Harwood, 1974).

differences in yield for individual border rows were noted.
These differences, although not statistically significant,
were also observed by Beets (1976). (See further Chapter
VII).

Other factors that may affect row direction are the slope
and length of the field. On land susceptible to erosion,
planting should be done along the contour, while on rectan-
gular fields, planting should be done length-wise in order
to minimize turning of machines or draft animals.

TIME OF PLANTING

The time of planting of crops depends primarily on agro-climatological factors during the season, the growing periods and growing cycles of crops. In countries of high latitude planting dates are dictated by temperatures while in most of the tropics, the planting date is dictated by rainfall.

In multiple cropping systems, the planting time is influenced by the factors mentioned above and by the growth cycles of the other crops grown on the same land during the year. The time of planting for sequential cropping systems generally depends on the moisture balance and the growth duration of the previous crop(s). The first crop in a rainfed double cropping system is normally planted as soon as the moisture balance allows. This crop must be early maturing, because the second crop will have to be planted before the rains tail off. If the rainy season is relatively short (5 to 6 months), the time of planting is often critical.

There are, however, additional factors governing the time of planting of sequential cropping systems. For example, destructive pests or diseases sometimes occur only during certain periods of the year and it is necessary to plant the crop when the damage will be minimal. An irrigated wheat/soya bean double cropping system practiced in Zimbabwe illustrates which factors can be involved and how they interact. In this country, the date of planting of wheat is governed by two factors a) susceptibility of the crop to rust; b) sensitivity of the crop to frost. In the area where this cropping system is practiced, temperatures just prior to planting are relatively high (maxima of 27 °C in April). When the wheat is planted too early, heavy .attacks of rust can be expected since the disease is most prevalent during hot weather. For this reason, planting is not done before May 15. When the planting is too late, however, flowering of the crop may coincide with early morning frosts in July and August. Extensive date-of-planting trials have shown that planting in the two last weeks of May is significantly superior to planting before or after this period. If soya beans, which are grown in summer, are planted too early, the ripening period of the crop coincides with the tail-end of the rainy season which results in crop losses and harvesting difficulties due to excessive moisture. Because there is little time available for land preparation between harvesting soya beans and planting wheat, minimum tillage techniques are employed.

In mixed and relay cropping, the sensitivity of plants to competition during the life-cycle of the species in the association must be considered. Many crops have clearly defined periods of high sensitivity, and stress during such periods influences the further development and yield of the crop. Cereals are usually sensitive during tillering and most other crops are sensitive in the transition period between vegetative and generative development. It is important to be aware of the stress periods of the species used in the association and plant them so that competitive effects are minimized during the stress periods of each species. This can be achieved by manupulating the "relative time of planting" of the components of the association: e.g., species A can be planted before B, simultaneously with B or after B. Generally, the earlier a species is planted, relative to the other, the less competition it suffers from the other species. This is illustrated with two examples from IRRI (1972).

Five crops were relay-planted with rice at four different times. In all cases but one the yields of the relay-crops were reduced when these crops were planted seven days later than the rice. As the time of planting of the relay-crop was delayed, the yields were reduced. The yield of maize (a crop which is very sensitive to competition) was less than one-third when the crop was planted three weeks after the rice, as compared to when the two crops were planted simultaneously. Yield reductions were smallest for cowpeas and sorghum. For cowpeas this is not surprising, since it is well known that the crop is highly shade tolerant.

In another trial, maize was planted up to 80 days after soya beans. The soya bean yield increased dramatically as the maize planting was delayed up to 20 days. With 20 to 60 days delay the soya bean yield decreased slightly and when the maize planting was delayed for more than 60 days, the soya bean yield increased again (IRRI, 1973).

Because the initial growth of soya beans is rather slow it seems beneficial to delay the planting of maize for some weeks. As shown in Figure 5.6, however, a delay of maize planting of three weeks or more will result in severe competition for the maize.

Another example of associations, where relative time of planting influenced the yield of crops, was given by Evans and Sreedharan (1961). Mixtures of castor-bean with groundnuts and soya beans were planted. It was found that the absolute date-of-planting had no effect on yields of castor-beans in pure stands, but when intercropped with groundnuts or soya beans there were significant yield reductions for both crops at the later planting dates.

FERTILIZATION

In sequential cropping systems, the nature of a first crop and the fertilizers used are likely to affect the performance of the second crop. Similarly, the first crop can have either a beneficial or a detrimental effect on the second crop. The magnitude of the effects varies considerably but generally they are not great. The greatest effect is on the availability of nitrogen. It is, however, difficult to generalize since the residual effect of nitrogen fertilization is affected by many variables. Benclove

Figure 5.6 Effect of relative time of planting of the components of a maize/soya bean association. The maize
was planted 20 days after the soya beans and is
heavy shaded and suppressed.

(1970) found that a groundnut crop, if ploughed under after the harvest of the nuts, can return to the soil 32 kg
of nitrogen and 25 kg. potassium per hectare. Hence, if
groundnuts and maize are grown in a double cropping system,
the nitrogen fertilizer for the maize following groundnuts
can be reduced. In Malaysia, where the above was found,
maize can only be successfully grown with heavy applications of lime and phosphate. If groundnuts follow maize,
then the legume may be able to utilize the residual effect
of the lime and phosphate applied to the maize. Reddi, et
al (1973) found that the residual effect of nitrogen
application to rice on a succeeding soya bean crop
increased in yields from 1.3 to 1.9 tons/hectare when the
nitrogen applications to the preceding rice crop increased
from zero to 180 kg. Nitrogen/hectare. While residual
effects are site-specific, it appears that some residual
effects can always be expected and should be considered in
fertilizing succeeding crops.

Adequately managed and fertilized sequential cropping
systems can maintain production almost indefinately. This
is illustrated by the work of Lin, et al (1973) who double
cropped rice in Taiwan for 48 years. Average rice yields
were constant for adequately fertilized treatments (2.5
tons/hectare/crop); unfertilized treatments gave also
similar yields but at lower levels (1.6 tons/hectare/crop).

In mixed cropping, fertilizer response of the individual

species may change drastically because of interference from the other species in the association. Since the nutrient requirements of the components of the association partly overlap, it is not sufficient to apply fertilizers according to the needs of the individual crops, or to sum the fertilizer requirements for the components. There are complicated interactions which are difficult to measure. The effect which an intercrop has on the main crop depends on characteristics of the crop such as growth cycle, nutrient requirements and the competitive power of the species during certain periods in the growth cycle. This is illustrated by the work of Enyi (1973) who reported that intercropping maize with either beans or cowpeas had more adverse effects than pigeon peas on maize yields. This was attributed to the fact that high rates of nutrient absorption by the two legumes coincided with the uptake by the maize crop, whereas with pigeon peas, the greatest nutrient demand occurred after the maize crop had been harvested. Hao (1972) used radio tracers to evaluate competition for nutrients between sugarcane and intercrops and found that groundnuts were less affected by the sugarcane than sweet potato.

In order to determine the exact fertilizer requirements of associations, extensive research using sophisticated experimental designs is necessary. However, when limited experimental data is available, and an association of a high fertilizer demanding cereal and a low fertilizer responsive leguminous crop is grown, it seems justified to apply the minimum requirements of the legume as a broad-casted dressing and the full requirements of the cereal as a side dressing.

IRRIGATION

In semi-arid regions, sequential cropping can only be practised with complete irrigation. Sequential cropping has little influence on the irrigation method used, although in some cases overhead systems may be preferable to flood irrigation, since the land preparation for flood irrigation generally takes more time, which may result in excessively long turn-around periods. In situations of adequately assured water supply, maximum crop production per unit area is the objective. This can be achieved by planting at the optimum times which is often earlier than is practised by the average farmer. Irrigation should cease well before maturity of the crop, so that land preparation for the following crop is not delayed. Often, residual moisture can facilitate tillage operations.

The water requirements for crops grown under conditions of unlimited availability of water depend on evapotranspiration and rainfall. As shown below, studies done at Kharagpur, India by Mittra and Pande (1972) indicate that total water requirements of triple cropping systems vary considerably:

Cropping System	Water requirements in mm.
Rice-Rice-Rice	3449
Rice-Wheat-Rice	2543
Rice-Potato-Maize	2060
Rice-Wheat-Jute	1957

In the experiment, the moisture was partly supplied by rain and partly by irrigation (respectively 73, 70, 69 and 76% of total requirements).

Where water supply is limited and seasonal in nature, it should be used as efficiently as possible. When water can be stored, taking into consideration storage losses due to percolation and evaporation, the water requirements of the first crop need to be met only to the extent of roughly two-thirds to three-quarter and water thus saved can be diverted to the subsequent crop(s). For the first crop, water should only be applied at critical stages in its development. Moisture stress during germination, flowering and grain formation or other sensitive stages, may lead to severe reductions in yields. The next crop in the rotation should receive at least 200-300 mm of water in order to produce a minimum yield. Crops best suited for these circumstances are deep rooted and drought resistant ones.

In areas with a rainy season of between 120 and 150 days, rain-fed, double cropping is often not feasible due to a shortage of moisture. With small quantities of supplementary irrigation, however, double cropping can sometimes be done by planting the second crop as soon as possible after the harvest of the first, at the end of the rainy season. This crop will initially use residual moisture stored in the soil and later some supplementary irrigation is needed. This supplementary irrigation often does not have to exceed 200 mm. Generally, one crop per year grown under a high moisture regime will give a lower total yield than when two crops are grown, dividing the available water between them.

Little is known of the water requirements of mixed cropping systems. Because the evapotranspiration of a cropped area is dependent mainly on the evaporative demand of the climate, the quantity of water required to permit potential growth and yield of any one crop or a number of crops grown in a given area would remain nearly constant, irrespective of the number of species. For example, when a tall statured crop, requiring wide spacing, is grown, the evaporation from the soil exceeds the transpiration from the plants in the initial stages of growth. When the tall statured crop is interplanted with a low growing species, the evapotranspiration of the association is made up of the transpiration of the two species plus the evaporation from the soil. In the case of an association there will be a greater leaf area resulting in a higher rate of transpiration, but also more shading of the soil, thus reducing evaporation. Hence, when changing from monocultures to polycultures, there is a shift from evaporation from the soil to transpiration from the leaves with the total evapotranspiration remaining more or less constant. Consequently, total water and irrigation requirements of monocultures and polycultures do generally not differ much. Water requirements during certain periods in the growing period may, however, be different. When the periods of high moisture requirement of two crops coincide, and if this happens at times that moisture stress seriously reduces yields, it may be necessary to irrigate relatively large amounts of water.

EROSION

The rate of erosion of a cropped field depends on five factors:

    (i)   rainfall;
   (ii)   soil erodability;
  (iii)   length of slope;
   (iv)   slope gradient; and
    (v)   vegetative cover.

The last factor is especially important when considering multiple cropping. The better the soil cover, the less the erosion. The soil cover provided by different field crops varies. Intertilled crops tend to encourage erosion. For example, maize, which is one of the most common tropical crops, generally provides a poor soil cover, and erosion is consequently often a problem. When maize is mixed-cropped with a second species, especially a legume, erosion hazards are reduced because overall soil cover is improved.

Strip cropping is widely advocated in the United States to combat erosion (Brady, 1974). The strips consists of a number of rows of an erosion-susceptible crop alternated with rows of a crop that limits soil loss. The width of the strips depend upon the degree of slope. Practical widths vary from 30 metres for a slope of 5 per cent to 15 meters for a slope of 20 per cent (Wischmeier and Smith, 1965).

CROP MANAGEMENT

Sequential cropping systems require high standards of management. What is particularly important in these systems is the length of the turn-around period (time between harvest and planting of subsequent crops). Short turn-around periods can only be achieved if the soil condition immediately after the harvest of a crop is suitable for cultivation. The soil should not be too wet or to weedy and there should not be an excess of plant debris. Therefore, standards of weed management have to be high throughout the life of a sequential cropping system.

It is, however, often not possible to achieve short turnaround periods without mechanization. Similarly, chemical weed control often has to be introduced in sequential cropping systems.

In mixed cropping, the most important factors that determine the level of management are "weeding" and "fertilization". Whereas most mixed cropping systems practised by tropical farmers are generally badly weeded, mixed cropping trials conducted by research workers are often done under a high level of weed management, and fertilizers are almost always applied.

Under indigenous conditions, where crops are inadequately weeded, yield losses due to competition from weeds are often greater in monocultures than in associations since the good soilcover provided by mixed cropping reduces the need for weeding because weeds are then heavily shaded and killed under low levels of light. This plant-weed competition is influenced, not only by the crop species in the association, but also by fertility levels. Nitrogen has the greatest influence on this competition. An example of actual weed response interaction with a crop and nitrogen level is shown in Figure 5.7. Weed growth did not increase significantly under maize as the nitrogen level was increased. The increase with mung beans was slight, but groundnuts failed to suppress weeds at high fertility levels and 3.4 tons/hectare of weeds resulted. In all crop associations weed growth was less than in comparable monocultures.

The weather, and especially light conditions, during the growth period influences crop-weed-fertility interactions. Wet weather and low light intensitites generally reduces the growth of the intercrop, but weeds (depending on the species) are often more shade tolerant, and in this case the intercrop is unable to suppress the weeds.

When fertilizer responsive high yielding varieties are used, crop associations generally respond to high levels of management. (Palada and Harwood, 1974; Beets, 1977). At low management levels, however, mixed cropping systems generally perform better than monocultures because mixed crops suppress weeds better than sole crops.

MECHANIZATION

Mechanization is often considered a prerequisite for sequential cropping systems. To achieve higher cropping intensities and yields per unit area, timely performance of farm operations from land preparation to harvesting and

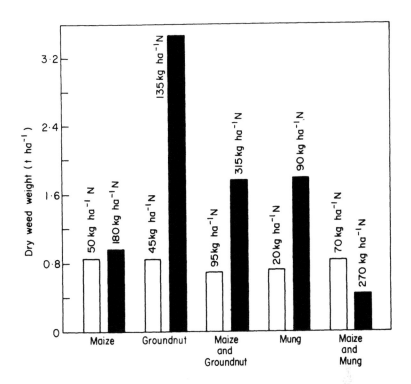

Figure 5.7 Interaction effects of crop combinations, weed
control and fertilizer level on weed growth.
(From IRRI, 1973).

marketing are important. Often, intensive agriculture and
tight rotational schedules leave little or no time gap
between the crops, and operations like harvesting, post-
harvest handling and land preparation must be mechanized.

There are no special problems in mechanizing sequential
cropping since conventional implements can be used for all
operations.

Mechanization for mixed cropping systems is difficult and
this is often cited as one of the main reasons against
these systems. Mechanization is, however, often neither
necessary nor desirable, since mixed cropping practices are
frequently found in areas with labour surpluses. In these
areas it is often desirable to use labour-intensive
production methods, rather than labour saving, mechanized
techniques. While seedbed preparation for mixed cropping
systems can generally be done mechanically usually cultivation, spraying and harvesting can only be done manually.

The adoption of mixed cropping in the technologically
advanced countries is hindered by the difficulty of
mechanizing these systems. As agricultural machinery
becomes more sophisticated, this disadvantage may become
less important. In the future, it may be possible to

design machines with electronic and hydraulic devices that can differentiate between different plant species. Such machines could be used for weeding and harvesting crops that are grown in mixtures.

VI Plant interrelationships and competition

INTRODUCTION

When plants are grown together in a community, they will affect each other. There will be "interference" and the result of this interference can be called "interference effect". (Lampeter, 1960). Interference occurs between plants of the same species, between plants of different species and also between different parts of one plant; e.g. between shoots and fruits of the same plant. The following terminology can be used when describing interference:

    (i)   Intraspecific:among individual plants of the same species;

   (ii)   Interspecific: between plants of different species; and

  (iii)   Interplant: between parts of a single plant.

The nature and effect of interference is of great interest since it has bearing on almost all processes in the individual plant as well as on the "plant community" or "crop". Interference will frequently occur in the form of "competition". Competition is a physical process. With few exceptions, such as the crowding of tuberous plants when grown too closely, an actual struggle between competing plants never occurs. Competition arises from the reaction of one plant upon the physical factors about it and the effect of the modified factors upon its competitors. Two plants, no matter how close, do not compete with each other so long as the water content, the nutrient material, the light and the radiation are in excess of the needs of both. When the immediate supply of a single necessary factor falls below the combined demands of the plants, competition begins. Since the environmental resources necessary for growth are usually in limited supply, competition almost always takes place at some stage in the development of a plant community. The time at which competition will commence depends on:

    (i)   the level of supply of resources; e.g. soil fertility, radiation, moisture balance; and

   (ii)   the nature of the plant community and in particular the resource requirements of the individual plants, the number of plants per unit area (plant population) and spatial arrangements.

As plant populations of monocultures are increased, competition will commence earlier and will be more severe. It is more severe because the individual plants in the community all require the same resources at the same time. On the other hand, in mixtures, different species require different resources and competition is less likely to take place. The potential advantage of growing species in associations therefore depends primarily on the degree of INTER crop versus INTRA crop competition (resp. competition between plants of different and of the same species). This is usually studied in experiments which are set up as "replacement series".

REPLACEMENT SERIES

In replacement series the yields of different species are compared with their yields in monocultures by gradually replacing a species (a) by a monoculture of another species (b). A two-phase replacement series is done in two steps as follows:

100%	a	66%	a	33%	a	No	a
No	b	33%	b	66%	b	100%	b

	Phase 1	Phase 2	
Monoculture	Mixed Cultures of		Monoculture
of Species a	Species a and b		of Species b

If the species respond in the same way as in spacing experiments, the yield curves will be parabolic, or first linear and later asymptotic, as is shown in Figure 6.1. In this case, competition between the species does not occur. The two species apparently do not interfere with each other. Although the species grow close enough to affect each other, they seem to be indifferent to each other, which means that they occupy entirely different "spaces". (Space, defined by v.d. Bergh (1975) as the integrate effect of all biotic factors on the growth of a species). This situation normally does not occur. Generally, there is interference between the species, often in the form of "competition" but other ways of interference also occur.

EXISTING PLANT INTERRELATIONSHIPS

The ways of interference have been described by many authors and several terms have been proposed. The terminology used by v.d. Bergh (1975) is as follows:

(i) <u>Indifferent</u> - or Complementary (Trenbath, 1974) Supplementary or Independent

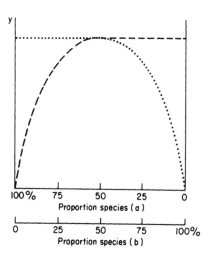

Figure 6.1   Replacement series of two species a (----) and
b (   ).

(Dalrymple,1971) or Neutral (Mather, 1974);
   (ii)   Competition - or Mutually Harmful (Mather,
         1974);
  (iii)   Hampering; and
   (iv)   Stimulation - or Complementary (Dalrymple,
         1971) or Mutually Beneficial (Mather, 1974).

A relationship is entirely "indifferent" when the plants do not interfere with each other. This does not normally occur, but would apply to double cropping with clearly separate seasons, or where adequate fertility and moisture are available for all species.

When two species compete, a yield increase of the one species results in a yield decrease of the other species. When the yield decrease is equal to the increase there is "pure competition". The resource requirements of the species are exactly the same, or, they occupy and compete for exactly the same "space".

Hampering effects are normally the result of toxic secretions of one of the species in the community. Cowpeas is an example of a crop that secretes a substance which is harmful to other plants.

Stimulation occurs when the productivity of a species is increased by some action of another species. The best example is the excretion of nitrogen by a leguminous plant and the uptake of this nitrogen by another species. Another example is when the micro-climate is changed by one species, resulting in it becoming more favourable for the other species. This occurs in annual windbreak cropping systems.

Relationships normally change during the development of a plant community. Frequently, the relationship will be indifferent at the early stage of vegetative development and will later, as the plants grow bigger and their requirements of "space" increase, become competitive. The nature of interference affects the growth and yields of crop associations.

RELATIVE YIELDS

In Figure 6.1 the yields of the two components of a mixture are equal in the monocultures. In practice, it is rarely possible to find crops or genotypes that give exactly the same yield. Frequently, the yield potential of the one species in the association will only be a fraction of the yield potential of the other species. Hence, the "replacement diagrams" will look like as illustrated in Figure 6.2.A.

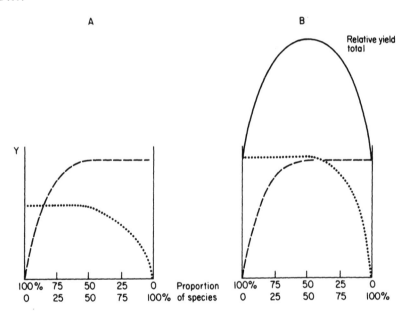

Figure 6.2 "Absolute" or real yields in weight per unit area of a replacement series of two species a and b and same yields converted to "Relative" yields and "Relative Yield Total" (———) (B).

For a better comparison of the peformance of the species, the "absolute yields" (Figure 6.2.A) can be converted into dimensionless "relative yields" (Figure 6.2.B). The relative yield of the species is the quotient of its yield in the mixture and of its yield in the monoculture. When two species are grown in association, the yield per unit area is the sum of the yields of the components of the association. It is convenient to sum the "relative yields" and not the "absolute yields". This sum is called the "Relative Yield Total" (RYT). In Figure 6.2.B the relative yields of both species in the 50/50 proportion are equal to 1 and the sum of the yields (RYT) is therefore equal to 2.

In Figure 6.3 all types of interference together with "absolute yields" converted to "relative yields" are given. The higher relative yield totals (2.25 and 2.0) are obtained in figures 6.3.d when the two species stimulate each other: e.g. a yield increase in one species will result in a yield increase in another species. In figure 6.3.a the two species do not affect each other and the Relative Yield Total is therefore the sum of the yields of the two components of the mixture. In Figure 6.3.b the two species compete with each other and an increase in yield of one species will therefore result in a decrease in yield of the other species. When the yield increase is equal to the yield decrease, the RYT is equal to the yield of one of the species grown on its own; namely 1.

The relationships between crops grown in multiple cropping systems can also be described in economic terms (Dalrymple, 1971) by using a concept of generalized output interrelationships. The pure forms of these relationships are presented geometrically in Figures 6.4.A-C. The solid lines represent the production possibilities. The dotted lines indicate the varying amounts of Z which would be produced as output of Y is expanded from P_1 to P_2, or a movement from A to B along the product possibility line. While the production possibilities have been presented in linear form, in reality, the relationships are likely to be curvilinear. Figure 6.4.A represents a Competitive relationship. Here, the output of one crop can be increased only through a drop in production of the other. Figure 6.4.B represents a Complementary relationship. Here, the output of one crop can be increased while the output of the other crop also increases. Figure 6.4.C represents a Supplementary relationship. In this case, the output of one crop can be increased without having any influence on the output of the other. These pure forms described above are unlikely to be found in reality. Combinations of each involving a competitive relationship are more likely to occur. This is illustrated in Figure 6.4.D where the increased output of one crop might initially result in some increase in output of the other, but beyond a certain point (B) the relationship becomes competitive. This relationship is Complementary-Competitive. In Figure 6.4.E the output of each crop will initially expand independently of the other, but beyond a certain point the relationship becomes competitive; later, when one crop ripens and its demands on the resources decreases, the relationship again becomes supplementary. This relationship is Supplementary- Competitive and is quite common since some resources become limiting only beyond a certain point. This will, for example, occur

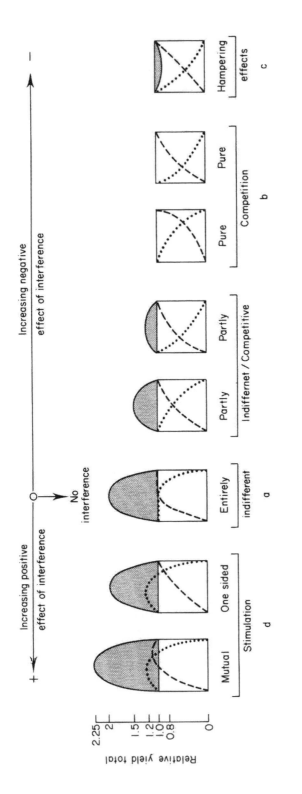

Figure 6.3 Diagrammatic presentation of the various types of interference. The solid line represents the overall performance, the remaining two lines indicate the performances of the individual components of the mixture (After v.d. Bergh, 1975).

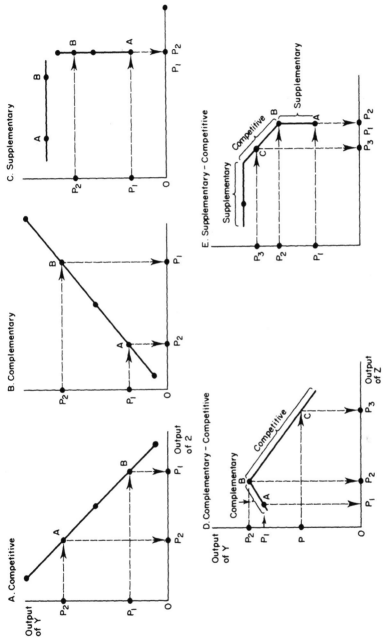

Figure 6.4 Production possibility lines (A-B and B-C) of two crops (Y and Z) involved in a multiple cropping system (From Dalrymple, 1971).

when moisture is abundantly available at the beginning of the growing period, but later becomes limiting and plants start to compete for this "input".

COMPETITION AND PLANT GROWTH

Above-ground interrelationships

The influence plant canopies of different species can have on each other can be divided into:

(i)   shelter effect (reduction in wind speed); and
(ii) shading effect (reduction in radiation flux).

The effects of shelter and the change in micro-climate it induces are discussed in Chapter VII. Shading effects are generally associated with competition for light since shading is frequently harmful.

When the canopy of one component of an association is set higher than that of another, the taller canopy (dominant species) intercepts the greater share of the light. If the soil conditions are non-limiting and the shaded species is not extremely shade-loving (which is rarely the case), the photosynthetic and growth rates of the shaded plants will be near to proportional to the radiation which they intercept (Stern and Donald, 1962; Santhirasegaram and Black, 1968). The plants will adapt to a certain degree to the low light levels but some reduction in yield usually results. The general conclusion from all experiments involving competition for light is that the component with its leaf area higher up in the canopy of the community is at an advantage. According to Stern and Donald (1962) it is also likely that, if the leaves are horizontal, the advantage is greater than if they are erect since horizontal leaves intercept more of the total downward light flux per unit area of leaf than do erect leaves. Since soil fertility as well as moisture balance can be relatively easily controlled, competition for light differs from that of nutrients or water in that there is no "common pool" from which plants can draw their supplies. Light energy is instantly available and it must be instantaneously intercepted and cannot be stored.

Below-ground interrelationships

In the early stage of plant development, roots of individual plants will be far enough apart from each other not to interfere with the supply of soil factors to its neighbours. However, since the surface area of the root system is very large, at some stage in the development of a crop, competition for supplies may begin. When the cropping system consists of different species, overlapping of root systems of the same species within the mixture is likely to begin earlier than for different species. Therefore, intra-specific competition is likely to start earlier than inter-specific competition. The degree of overlap between components' root systems determines the intensity of competition effects (Cable, 1969; Trenbath, 1975). The spatial distribution of individual roots, as well as whole rooting systems, will most likely influence the intensity of competition. Nelliat, et al (1974) studied the distribution of coconut roots and found that the vertical distribution of roots is such that the top 30 cm layer of the soil was practically devoid of functioning coconut roots whereas pineapple roots were found to be restricted to a depth of about 30 cm only. Hence, theoretically, it would be possible to have a mixed cropping system of coconut and pineapple without below-ground interspecific competition and, indeed, this crop combination is quite common.

Water uptake produces a gradient of water content around the roots. As the soil around the root is dried out, water will flow to the depleted soil. Depending on a series of factors such as hydraulic conductivity and water content of the soil, the depletion zone for water can extend up to 25 cm from a single root (Klute & Peters, 1969). This means that the depletion zone is fairly large and competition for water is expected to occur as soon as the depletion zones of roots of the different components of a crop association overlap. Competition for water is therefore closely linked with spatial arrangements and rooting patterns (Willey, et al, 1976). Rooting systems seem to avoid each other to prevent competition. This will often result in a deeper penetration of roots in crop associations which means that more water will be available and competition will be less than expected (Lakhani, 1976; Fisher, 1975).

When the species are grown in association the following factors determine the nature and extent of competition for water and nutrients (Barley, 1970; Bowen, 1973; Andrews and Newman, 1970):

    (i)   Root production. Early, fast penetration of
          the soil will often result in a competitive
          advantage;
    (ii) Root density;
    (iii) Proportion of the root system actively
          growing; and
    (iv) Water and nutrient uptake potential.

According to Kawano, et al (1974), early uptake seems to be the key to success in competition for mobile nutrients. The factor 'time' also plays an important role elsewhere, e.g. when the nutrient requirements of the various species occur at the same time, competitive effects can be expected to be larger than when the species take up the elements at different times. For example, Enui (1972) found that competition had a greater depressing effect on the growth of a cereal crop in case of an association with cowpeas than with pigeon peas. Since the cereal was in its reproductive stage at about the same time as the cowpeas, the depressing effects of this legume on the cereal might be attributed partly to the higher nutrient requirements of the former and partly to the fact that the period of high nutrient absorption by the legume coincided with that of the cereal. Flowering of the pigeon peas did not take place until the cereal had been harvested, so that in this association the period of greatest nutrient demand occurred when the cereal had completed its growth cycle.

Chang, et al (1969) studied the competition for nutrients between sugarcane and two intercrops (sweet potato and groundnuts) by following the recovery of fertilizer-applied P and K. Measurable effects of intercrops and fertilizer placement on recovery patterns for P32 and Rb 86 were observed. As illustrated in Figure 6.5 there were no differences on fertilizer P uptake between sugar cane and

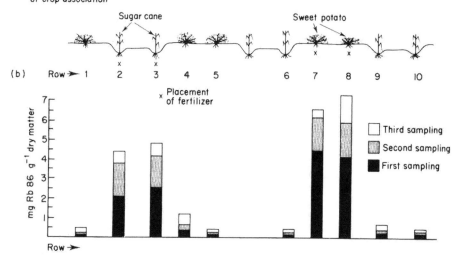

Figure 6.5 Some results of a competition trial with sugar-cane and sweet potato (After Chang, et al, 1969).

sweet potato but the amount of fertilizer Rb86 (or K) up-take by sweet potato was more than the uptake by sugarcane. Further, when the sugarcane was interplanted with ground-nuts, the absorption of the nutrient P32 or Rb86 by sugar-cane was more than by the groundnuts. This means, they postulated, that the groundnuts were less affected by the interplanting.

In another competition study, pearl millet and pigeon

peas were grown in various associations and it was found that there was root competition between pearl millet plants having lower root cation-exchange capacity for uptake of K and pigeon peas having relatively higher root cation-exchange capacity for uptake of Ca (Daftardar and Savand, 1971). A shift in the composition of the association resulted in a change in competition. Gray (1953) also found that the competition between a graminae and a legume for K is directly related to the cation-exchange capacity.

In mixtures of components adapted to soils of different nutrient status, the species or genotypes adapted to low nutrient soils have been found to be more agressive on such soils (v. d. Bergh and Elberse, 1962). When nutrients are added in such situations, the competitive power of the species will change, or, in other words, the relative agressiveness of a genotype in a given mixture varies greatly from crop to crop in response to environmental conditions (Trenbath, 1974).

Competition for nitrogen is discussed separately since this element is more mobile, and plays a very important role in plant production. Nitrate in the soil is in the form of mobile ions and is carried passively in moving water. The nitrogen depletion zones will therefore be as large as those for water, provided the ions are taken up as fast as they arrive at the roots (Barley, 1970). The mobility of nitrogen together with a great need for it by most plants may lead to severe competition for this element. However, there could be one exception: when a leguminous plant is grown in association with a non-legume since leguminous plants can fix their own nitrogen from the air. However, legumes can readily use either symbiotic or combined nitrogen. But the amount of symbiotic nitrogen produced is inversely related to the amount of combined nitrogen available. When supplied in excess of amounts needed for plant growth, combined nitrogen may prevent symbiotic fixation (Hinson, K. 1975). This might mean that legumes do compete for nitrogen when grown in association. The results of studies conducted by Beets (1976) which are discussed in Chapter VII, support this.

From the experimental evidence available it cannot yet be concluded to which extent a leguminous plant competes for nitrogen with a non-legume with which it is planted in association. Neither can it be concluded that leguminous plants do, or do not, fix nitrogen which will later become available to another plant in the association. In general, the interrelationships are ill understood and more experimental work is required. However, it can be concluded that the following factors are likely to affect the level of competition between a legume and a non-legume:

  (i)   The level of available nitrogen in the soil;
  (ii) The ability of the legume to fix nitrogen. This will depend on the species and the azotobacter strains in the soils;
  (iii) The light intensity; and

(iv) The time of overlap of the species in the association.

COMPETITION AND CROPPING SYSTEMS

The influence different species have on each other depends primarily on the botanical characteristics of the species. Normally, one species will suffer more than the other when grown in associations. Or, one species may be more successful than the other in securing an undue "share" of the "space", i.e., the light, the water or the nutrients, and as a consequence its yield per plant will be only slightly decreased, not affected, or even increased. In which case we have a situation of "dominance" and "suppression" or an "aggressor" and a "suppressed" or "subordinate" species. This relationship is shown diagramatically in Figure 6.6. Species A is the aggressor and yields 6 weight units per

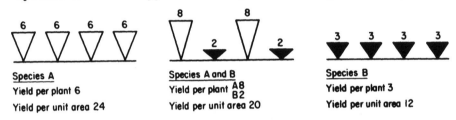

Species A
Yield per plant 6
Yield per unit area 24

Species A and B
Yield per plant A8
B2
Yield per unit area 20

Species B
Yield per plant 3
Yield per unit area 12

Figure 6.6 Diagrammatic presentation of the yield relationships commonly found when two species are grown separately and in association at "normal" seed rates. (From Donald, 1963).

plant and 24 weight units per unit area when grown as a monoculture. When grown in a mixed culture, it yields more (8 weight units per plant), because the plants are wider spaced and the plants of species B apparently compete less with species A than plants of species A compete with each other. (Intra- specific competition is larger than inter-specific competition). However, species B yields more in a monoculture than in a mixed culture (resp. 3 and 2 weight units) because in the latter it is suppressed by species A. This situation often occurs with a tall and a short plant and with a legume and a cereal.

Papadakis (1941) made extensive comparisons of cereal/legume mixtures for grain production and he found that the cereal grain produced by a 1 hectare mixture was 61 per cent more than the grain produced by 1/2 hectare of the cereal grown alone. On the other hand, the grain of the leguminous plant produced by 1 hectare of the mixture was 9 per cent less than that produced by 1/2 hectare of the legume grown alone. The total yield was 21 per cent higher than the average of the yields of the two plants grown alone.

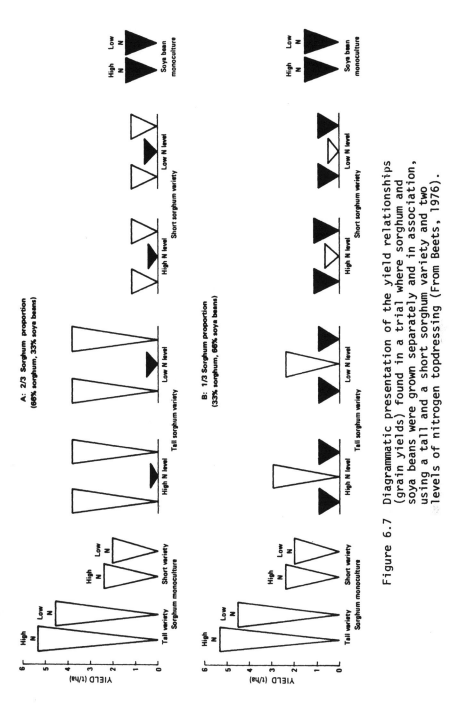

Figure 6.7 Diagrammatic presentation of the yield relationships (grain yields) found in a trial where sorghum and soya beans were grown separately and in association, using a tall and a short sorghum variety and two levels of nitrogen topdressing (From Beets, 1976).

Beets (1976) obtained similar results when he grew two varieties of sorghum together with soya beans in a two-phase replacement series (see Figure 6.7). The two varieties of sorghum were a short and a tall variety. Two levels of nitrogen top-dressing were used. In both phases of the replacement series, a higher soya bean yield was obtained when the legume was grown in association with the short sorghum variety. In the one-third soya bean proportion the yield of soya beans was lower when the sorghum was topdressed; 25 per cent in case of the short sorghum and 33 per cent in case of the tall sorghum variety. The short sorghum variety suffered more from competition from the soya beans than the tall variety.

In the above experiment only two factors i.e. plant height and nitrogen level were studied. However, other factors such as plant densities, moisture balance, relative time of planting of species, all influence the outcome of a crop association.

VII Agro-ecological, biological and plant physical aspects

CLIMATE AND SOIL

Introduction

The tropics are characterized by a rather regular climate with regard to solar radiation, air temperature, wind speed and evaporation. Annual air temperature fluctuations generally have only a marginal effect on plant growth since temperatures are normally conducive for plant growth throughout the year. In most regions of the tropics, and especially in the semi-arid regions, water availability is the major constraint to agriculture, particularly to year-round agriculture. Therefore, cropping systems normally reflect local moisture conditions. When considering moisture availability in relation to cropping systems, the relevant parameters are:

  (i)   seasonal rainfall regimes;
  (ii) intensity and effectiveness of rainfall;
  (iii) variability and reliability of rainfall; and
  (iv) evaporative demand.

Seasonal rainfall regimes

In the tropics, seasonal rainfall patterns can often be related to farming systems and problems of water supply. There are two broad categories: uni-modal and bi-modal rainfall patterns. Generally speaking, bi-modal patterns offer the largest scope for sequential cropping (under rainfed conditions) since there is enough moisture for more than one crop. Uni-modal rainfall patterns which are of sufficient duration (at least seven months) to support two or more consecutive crops, can only be found in limited regions in the world. Mixed and relay cropping systems' are often advantageous in areas with uni-modal rainfall patterns of relatively short duration because, in such situations, it is important to grow the maximum number of crops when adequate moisture is available.

In addition to the length of the rainy season(s), the "severity" of the dry season is important because it determines the extent to which crops can survive during the dry period. Generally, when there is less than 100 mm of rain per month, the dry season is classified as severe and most crops cannot survive if the drought lasts longer than three months.

Areas with severe and long dry seasons and relatively short uni-modal rainfall patterns are found mostly in northern and southern Africa and the Indian sub-continent. Moving from these regions towards the equator there are areas with:

(i)   a single rainy season and a single dry season;

(ii)   a bi-modal rainy season separated by a relatively more pronounced dry season; and

(iii)   a bi-modal rainfall pattern with no severe dry season around the equator.

Intensity and effectiveness of rainfall

In the tropics, a high proportion of rainfall occurs in large storms of high intensity. This characteristic is important for both soil erosion and the effectiveness of rainfall. Effective rainfall in agricultural terms is that portion of the water entering the soil and remaining within root range. Effective rainfall is lower than total rainfall because of water losses due to:

(i)   deep percolation;

(ii)   run-off;

(iii)   evaporation; and

(iv)   low waterholding capacity of the soil.

All these factors are, to some extent, influenced by cropping systems and crop management. There is a significant interaction between cropping systems and effectiveness of rainfall. Multiple cropping systems provide good soil cover which reduces run-off and evaporation. Both the effectiveness of rainfall and water-use efficiency are, therefore, often better with well designed multiple cropping systems than with monocultures with longer periods of partial soil cover. This is illustrated by work done at the Indian Agricultural Research Institute (1972) where field infiltration rates of four multiple cropping systems were measured in situ. As illustrated in Figure 7.1, it was found that the infiltration rates increased as the cropping intensity and the soil cover increased.

High infiltration rates not only increase the water-use efficiency but also result in less erosion because there is less run-off.

Downpours of high intensity can cause mechanical damage to plants, especially ·during those periods when the crop is sensitive to damage (e.g. tobacco when it is almost ripe and cotton when the bolls have opened). On the other hand, low intensity rainfall of long duration is usually accompanied by prolonged periods of high relative moisture content of the air. This condition encourages the development of

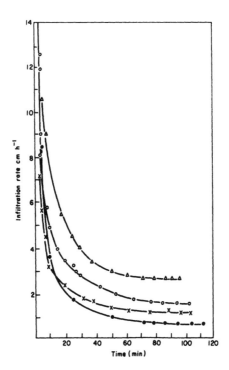

Figure 7.1 Field infiltration rates of four multiple cropping
systems in India.
(•) Double cropping of maize and wheat;
(x) Triple cropping of maize, mung beans and
wheat without cultivation and
(o) the same system with cultivation;
(Δ) Relay-cropping with mung beans, maize, toria
and wheat.
(From Indian Agric. Res. Inst., 1972)

fungal and other diseases. When planning cropping systems,
the intensity of rainfall during different seasons should,
therefore, be taken into account to help choose suitable
crops and planting dates.

Variability and reliability of rainfall

In areas with marked seasonal rainfall patterns, variability and reliability at the start and finish of the rainy
period are particularly important since the first determines planting dates and the latter determines whether
early or late maturing varieties should be used. Reliable

rainfall is defined as a specified amount of rainfall with a specified minimum probability of occurrence.

The amount of reliable rainfall in the pre-rainy period is small in many areas. This period is, however, important, since early planting is normally advantageous. It is, therefore, desirable that the probability of receiving a minimum amount of rain in this period is known. The minimum amount that is necessary to "start" the season varies from crop to crop and area to area. The best criterion to define the start of the season is whether the amount of rain during this period is sufficient to support plant growth.

The broad concept of "effective planting rain" can be used to roughly indicate the start of the season. "An effective planting rain" and the "effective start of the season" can be defined as follows:

(i)   the rain wets the top 5 cm of the soil to field capacity; and
(ii) no day of zero available moisture occurs in the 10 days following the planting rain.

Evaporative demand

Potential evaporation is more constant from year to year than rainfall because of the small variation in determinants such as solar radiation. Both rainfall and evaporation determine the availability of water which is the main factor governing cropping systems in the tropics. Cropping systems can, however, influence the actual evapotranspiration of the different components of the cropping systems and the soil evaporation/crop transpiration ratio, and hence consumptive water-use. Water consumption of a crop is defined as the sum of the water evaporated from the ground surface and that transpired by the crop canopy during the growing period. Inundated fields not covered by a crop, vegetation, or other soil cover, have a high rate of evaporation, especially during periods of high radiation intensity and strong winds. When a crop covers the soil, evaporation is reduced. Hence, the plant transpiration/ soil evaporation ratio is a function of leaf area which, in turn, depends on the cropping system. As this ratio increases, efficiency of water-use increases and conditions for plant growth improve. The influence of cropping systems on evaporation and micro-climate are, however, not usually highly significant.

WATER AND CROPPING SYSTEMS

Periods of water availability

"Availability of water" and especially "non-availability of

water" or "drought" should be expressed in terms of plant response to the moisture balance. Generally, definitions of availability of water which cover only rainfall are not satisfactory because plant growth, the ultimate objective, is not included as a criterion of moisture availability. Different environments can be compared in terms of plant growth potential. Also, both the suitability of an environment for crop production and the relative importance of different factors of a particular environment on crop growth can be assessed. In tropical agriculture "availability of moisture" is usually the most important criterion for crop production. When moisture is related to crop growth and production, there is a significant correlation between individual parameters such as total seasonal rainfall, frequency and length of drought, date of planting, and humidity and temperature. These correlations alone, however, offer little prospect of assessing the suitability of a cropping system to a particular environment, or, of estimating the long-term production of a particular cropping system in a specific environment, or, the relative merits of different crops in a particular environment. It is necessary to combine these individual parameters. The combination of the different ways of expressing moisture in terms of plant requirements (rainfall, drought, transpiration, evaporation) into the single concept of "moisture balance", will give better correlation with cropping systems performance than can be obtained when an attempt is made to correlate the separate factors with plant production. Cochemé (1968) gives an example of describing the "moisture balance" and "availability of water periods" of a semi-arid area in West Africa. Curves representing R (rainfall), E_t (Potential Evapotranspiration) were plotted together with fragments of curves representing $E_t/10$ and $E_t/2$. (See Figure 7.2). The two points of intersection $R=E_t/2$ define the boundaries of a period called "humid", during which there is a water surplus. The first two $R=E_t/2$ points delimit a "moist" period. A third defines the end of the "moist + reserve period", when up to 100 mm of water are stored in the soil at the end of the humid period. The length of this period depends on soil moisture storage characteristics. Two intermediate periods can also be recognized, one before and one after the humid period, which identify the moist period. The period for land preparation covers the period from $R=E_t/10$ to the beginning of the moist period.

When periods of "water availability" are described as above, a cropping calendar can be designed. When doing so the first point to consider is the length of the period for land preparation. The slope of the rainfall curve in this period will differ in various locations. This is important because the length of the period for preparation is directly related to the slope of the rainfall curve - the steeper the curve the shorter is the time available for preparation. The shorter this period, the larger the machinery pool or the bigger the labour force required, which, may be a constraint. The seriousness of this constraint may be

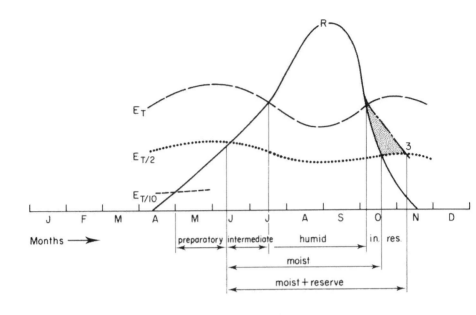

Figure 7.2   Graphical comparison of rainfall R with Potential
Evapotranspiration (E_t). $E_t/2$ and $E_t/10$ define
preparatory, intermediate, moist, moist + reserve
(res.) and humid periods.
(From Cochemé, 1968).

related to the structure and texture of the soil. Generally,
light reddish soils are easier to work than heavy tropical
black earths, especially when soil moisture conditions are
not ideal. Thus, they generally require less labour and/or
mechanization.

At the beginning of the moist period, when $Et/2=R$, actual
evapotranspiration of partially bare soil is about one-half
of potential evapotranspiration. Because water does not
limit crop growth, this time is suitable for sowing. The
length of the "humid" and "moist + reserve period" determines the most suitable cropping system. If the latter
period is less than about 90 days, sole and mixed cropping
should be practiced. If it is between 90 and 160 days,
relay cropping is the most appropriate and if it exceeds
160 days, sequential cropping is possible.

The next point to consider is matching the growth and
biological characteristics of the individual crops and
cropping systems. The major relevant biological characteristics are:

    (i)   length of growing cycle(s) of the crop(s);
   (ii)   occurence of periods of sensitivity to
           drought during the growing cycle and over-

all water requirements during the growing period;

(iii) sensitivity to wetness during the growing cycle, (e.g. many crops should be harvested in a dry period; when early maturing groundnuts ripen in a moist period, "sprouting in the field" will lower yields); and

(iv) sensitivity to pests and diseases (many crops are more sensitive to pest and diseases during periods of extreme drought or wetness; e.g., sorghum is more sensitive to fungi during wet than dry periods).

When these climatic and biological factors are taken into consideration, suitable crop species and cropping systems can be selected.

Actual water use

Water consumption of crops depends on plant characteristics and varies between varieties and environments. Kung (1971) estimated the average total water consumption for a number of crops in some Asian countries as follows:

	Water Consumption per month	Growing Period	Total Water Consumption
Lowland Rice	150 - 200 mm	5 months	750 - 1000 mm
Maize	85 - 100 mm	4 months	350 - 400 mm
Groundnuts	80 - 100 mm	5 months	400 - 500 mm
Soya beans	75 - 100 mm	3.4 months	300 - 350 mm

Because water consumption for individual crops differs, it is important to consider this factor when matching crops and environments.

The next issue to be considered is the total water consumption of a multiple cropping system. Will the water consumption of a double cropping system of, for example, rice and soya beans, be the sum of the consumption of the individual crops or will it be less or more? Similarly, what are the water requirements for a mixed cropping system? Few experiments addressing these issues have yet been undertaken. However, since multiple cropping systems provide better soil cover, less evaporation would be expected than for monocultures. Better soil cover also results in less run-off. The moisture retained in these ways can be used by the crops. Hence, it can be postulated that while water-use in the form of transpiration of multiple cropping systems is higher than in monocultures, evaporation and

run-off losses are less. Consequently, the environment may
have more moisture available (with equal inputs) for multiple cropping systems than for sole crops. This is illustrated by the work of Beets (1976) who measured the moisture
content of soils at successive depths under a maize monoculture and a maize-soya bean association. He found that
at all times and at all soil depths - with one exception in
each case - the soil of the association was dryer than the
monoculture soil indicating higher water-use by the association. (See Figure 7.3) If the season is dry, and mois-

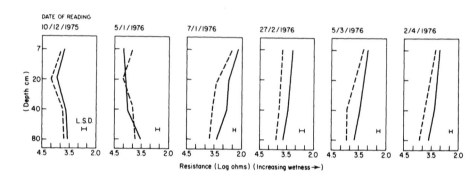

Figure 7.3   Measuring of soil wetness by nylon block resist-
             ances at successive depths in the soil under a
             maize monoculture (——) and a maize/soya bean
             mixed cropping system (---) at six occassions
             during the growing period. (Beets, 1976)

ture limits crop growth, the association could suffer more
from moisture stress than the monoculture (subject to
inter- and intra-plant competition and rooting patterns).
When water is in abundance, however, and this was the case
in the season these crops were grown, the increased moist-
ure-use increased growth and yields. In the experiment,
only at one date and at two depths was the soil of the
maize monoculture drier than the soil of the association.
This could possibly be explained by the fact that weeding
had taken place just prior to this moisture measurement and
that the stirring of soil increased the soil surface, which
resulted in a higher evaporation from the stirred,uncovered
topsoil of the maize monoculture than from the unstirred
soil covered (by soya beans) under the association. In
other words, the water evaporated by the soil in the monoculture was "lost water".

Crops differ in their reaction to moisture stress or drought, and a cropping system that is drought resistant in one environment may be quite unable to tolerate a less severe drought typical of a different environment. Important factors that affect the drought resistance of cropping systems are:

(i)   soil fertility;
(ii)   root development;
(iii)   plant population;
(iv)   shading or sheltering;
(v)   sensitivity of crops to drought at different stages; and
(vi)   time of planting.

High soil fertility normally enhances plant growth which generally means good root development. The better the root system the more easily the plant can extract water and the less susceptible it is to drought. In mixed cropping, root systems are often better developed than in monoculture systems which means that mixed cropping systems could be more drought resistant. (Subject to inter- and intra-plant competition).

Mixed cropping systems can also be less subject to damage from drought because the absolute plant population of the individual components of the association is lower than for monocultures. With lower plant populations there is more water available per plant and the risk of moisture stress is reduced. This is supported by Andrews (1973) who found that sorghum grown in association with cowpeas was less susceptible to moisture stress than sole cropped sorghum planted at a high plant population. This resulted from two factors: (i) the sorghum plant population was lower in the association than in the sole culture; and (ii) the sorghum was deeper rooted than the cowpeas and therefore did not suffer from moisture competition from the cowpeas.

One component of a crop association sometimes changes the micro-climate for the other component through shade or shelter reducing the evaporative demands of a crop which then needs less water to maintain turgor pressure. In such systems, the sheltered species suffers less from drought.

The quantitative importance of drought in crop production depends on the time and stage of development when the drought has the greatest impact on yield. Some crops have an all-round resistance to drought while others are relatively resistant only during certain stages of development. Maize, for example, is quite drought resistant in the seedling stage since the plant is protected by the first leaves which envelop it. Later, however, during the flowering (tasseling) stage the crop is very sensitive to drought. This also applies to most other cereal crops. Soya beans and most other legumes are sensitive to moisture stress during the pod formation and pod filling stages.

In the design of cropping systems, it is essential to know the sensitivity of crops to drought and to define the quantitative effects of water deficiencies during the different stages of crop development on yield. Planting dates should be chosen in such a manner that the periods of water deficiency coincide with periods of drought tolerance. When crops are planted in association, relative planting dates should be chosen so that the critical stages of the different species do not coincide.

LIGHT AND CROPPING SYSTEMS

Solar radiation distribution

Solar radiation distribution is closely related to rainfall in the tropics. Although radiation maps are useful guides for assessing the agro-climatic potential of different regions, radiation is a less critical factor than rainfall.

Radiation levels are highest in the dry zones of the tropics (up to 200 kcal/sqcm/year in the Sahara). These levels are not excessively high for agriculture per se, but since they are almost invariably accompanied by unavailability of water, they are of little use for agriculture. In most of the areas of the tropics with more moisture, the annual radiation varies from 130 to 170 kcal/sqcm/year which is considerably higher than for temperate climates (80 to 140 kcal/sqcm/year). Because high radiation levels are accompanied by high temperatures this further enhances the agricultural potential of the tropics.

In most areas, especially those with pronounced wet and dry seasons, there is a marked annual variation in radiation receipt. Because dry season radiation is always higher than in the wet season, yield potentials are also higher. Since the availability of water is normally low during this period, plant growth is retarded. If, however, irrigation is available, yields of most crops are higher in the dry season than in the wet season. In case of rice there can be a difference of about 20 per cent.

Leaf area and light interception

Whether crops are able to use high radiation and light levels depends on the inherent ability of species and on the Leaf Area Index (LAI). As the number of leaves and their size increases, light absorption and the rate of dry matter production also increase. The optimum LAI depends on the crop species, the season and the light intensity. A higher LAI will generally lead to more photosynthesis and, therefore, the ideal foliar development of a crop would hypothetically be the immediate attainment of the optimum LAI upon crop emergence. Relay cropping systems approach

88

this ideal somewhat since the canopy of such systems is formed of plants of different species which are in different stages of development. At a given stage the LAI of species A may be optimum or just below optimum, while the LAI of species B, which has just emerged, is low. The LAI for the two species together may, however, be optimum and the two species will be able to capture most of the light effectively. At a later stage, the LAI of species A will be reduced from optimum to nil (when the crop is harvested) but the foliage of species B will rapidly replace it.

In mixed cropping systems the situation is similar since the combined leaf area of the species in the association is normally larger than in monocultures and the build-up of the LAI is more rapid. This is supported by work undertaken at IRRI (1975) where LAI's, photosynthetic efficiencies, and dry matter accumulation of maize and rice mono and mixed cultures were measured. The maximum LAI for maize was reached six weeks after seeding and the maximum LAI for rice was reached twelve weeks after seeding, or, after the maize was harvested. (See Figure 7.4). The maxi-

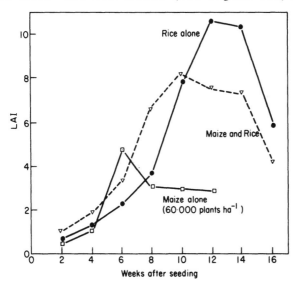

Figure 7.4   Total leaf area index of a rice/maize crop association (∇), a monoculture of rice ($\bullet$) and a monoculture of maize at 60.000 plants/ha ($\square$) over the growing period (After IRRI, 1975).

mum LAI for the association was between these dates. Maize alone had a relatively low leaf area duration (leaf area integrated over time), while rice alone had a considerably higher leaf area duration. As a result, the total dry matter production and grain yield of the associations was higher than those of either crop grown as monocultures.

With mixed communities, not only the combined LAI is important, but also the extent to which each species in the association contributes to the LAI. It is further important to know how much mutual overshading takes place by the species since a high LAI for a tall species may lead to excessive overshading of the lower species. Beets (1976) studied this in mixed cropping systems of maize and soya beans by measuring the "canopy cover" which is closely related to LAI. Figure 7.5 shows how the "per cent canopy

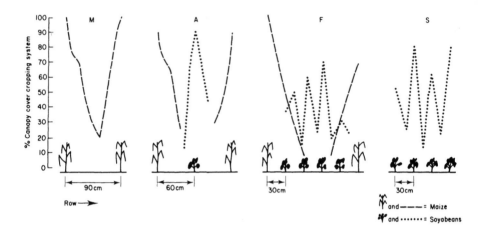

Figure 7.5   Percentage canopy cover in one monoculture of maize (M), one monoculture of soya beans (S), and two mixed cultures of maize and soya beans (A and F) (Beets, 1976).

cover" varies in space in monocultures (M) with a plant population of 44,444 plants per hectare and rows spaced 90 cm apart. Naturally, the canopy cover is highest just above the rows. At the time the measurements shown in Figure 7.5 were taken (five weeks after planting), the canopy cover just above the row was 100 per cent and the percentage cover gradually dropped to 20 per cent at the mid-point between two rows. Mixed cropping system A consisted of an alteration of maize and soya bean rows spaced 60 cm. The pattern of cover provided by the maize is similar to the pattern of the maize monoculture (M). Ninety per cent of the soil above the row of soya beans was covered but this cover quickly dropped to nil at about 20 cm from the middle of the soya bean row. There was no overshading of soya beans by the maize. In system F (three rows of soya beans between single rows of maize) the maize overshades the soya beans. As the LAI of the maize increases, so does overshading. The more intimate the association, the greater the rate at which this overshading takes place, and the greater

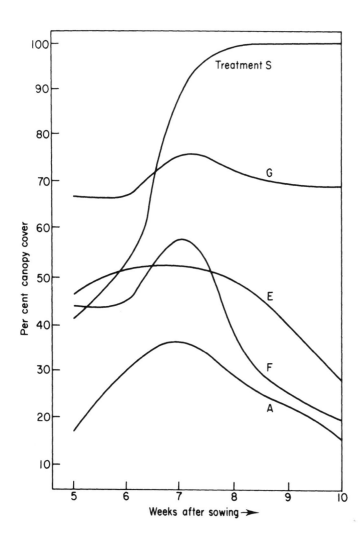

Figure 7.6   Percentage soya bean canopy cover as percentage
             of total canopy cover (maize + soya beans) in
             four mixed cultures. (Beets, 1976).

the proportion of the soya beans which is overshaded. This
is illustrated in Figure 7.6 where the soya bean canopy
cover is plotted as a percentage of the total canopy cover
(maize + soya beans), from planting to physiological maturity. In the monoculture system the canopy cover increases
to 100 per cent. In all mixed cropping systems, the soya
bean covers increases initially, but from the seventh week
after sowing all covers decrease,except for the system with
only a small proportion of maize. The decrease in cover is
due to overshading of maize and is greatest for the most
intimate system. The yield results of the trial show that
the yield performance of the mixed cropping system was

negatively correlated with the degree of overshading of the soya beans. Treatment G, where the soya beans were hardly overshaded, performed best with a Land Equivalent Ratio of 1:39. There was also a positive correlation between degree of overshading and lodging of the soya beans-in system G there was little or no lodging and the soya bean yield in system F was reduced because of severe lodging.

A reduced radiation level not only causes lodging, but also leads to changes in the physiological processes of the plant. Generally, shaded plants tend to grow taller and more spindly than solitary plants which results in an unfavourable grain/straw ratio. In monocultures grown with high plant populations, mutual overshading of leaves and lodging can be severe. On the other hand, when a tall crop is grown with lower populations in association with a low statured crop, the tall crop suffers less from intra-specific shading. Inter-specific shading does not occur since the shorter species is unable to affect the light environment of its neighbours. Under these conditions, plants do not need to grow tall to compete for sunlight and lodging of tall species is reduced.

Andrews (1973) found that neither millet nor maize grown in association lodged as the crops would normally do when grown as sole crops. Pendleton (1963) and Beets (1976) found that maize grown in strip cropping systems lodged less than in monoculture systems. Soya beans, the shorter species in the strip cropping system did, however, suffer from lodging.

Plant arrangement and light interception

Shading normally decreases yield either by reducing photosynthesis or by contributing to lodging. In order to minimize the reduction in light to a single plant, it seems necessary to maximize the distance between individual plants. This can be done by using equidistant planting patterns which minimize the competitive effects of neighbouring plants, thus maximizing yields. Donald (1963) showed that equidistant spacing gives highest yields in monoculture crops. Since the effects of plant competition play a greater role in mixed cultures than in monoculture the effect of planting pattern in the former is pronounced. In mixed cropping systems based on manual labour, it is feasible to plant crops in an equidistant pattern. As mechanization increases, however, and particularly in relay cropping systems, row cultures may be needed for management reasons.

When row planting is used, row spacing and row direction are important points to consider. The effect of varying the row spacing is relatively simple. Generally, the closer the rows, the more the pattern approaches the "ideal" equidistant pattern.

Row direction may also be of importance in multiple cropping systems. While few experiments have been done, the little evidence available shows that yields are greater from crops planted in north-south rows. This is not surprising since the light regimes in rows varies as the bearing of the row is changed. These effects seem greatest in strip cropping systems since differences in light regimes are more pronounced in such systems. Pendleton et al (1963) as well as Beets (1976) found that in strip cropping systems of maize and soya beans, the yields for the north soya bean rows in a strip planted east-west were considerably higher than for the south rows. Also, the yields for the east row in the north-south planting exceeded those for the west rows.

The row direction may also influence the photomorphogenic processes in mixed cropping systems.

The effects of plant arrangement and row direction may be summarized as follows:

  (i)   both inter- and intra-plant competition can be influenced;

  (ii) the light regime may be influenced through differences in light interception and shading; and

  (iii) the moisture regime may be influenced through differences in evaporative demands.

CROPPING SYSTEMS AND MICRO-CLIMATE

Introduction

The effects different crop species grown in association have on each other is often not direct but, rather, occurs because one specie changes the crop environment or micro-climate in such a way that growing conditions for the other species become more (or less) favourable. The environmental factors generally affected are light intensity and moisture availability. Light intensity directly affects the plant photosynthetic rate. The availability of water can also affect the photosynthetic rate because water is an essential component in the photosynthetic reaction. A shortage of soil moisture or atmospheric water causes stress on the plant and affects the efficiency of its photosynthetic reaction. The most direct influence of water availability on photosynthesis is its control of the stomatal aperture. As stomatas close, resistance to the diffusion of carbon dioxide increases.

Moss (1965) speculated on the influence of soil moisture stress and atmospheric demand on photosynthesis at varying light intensities as is illustrated in Figure 7.7.

When the soil moisture stress is increased, the optimum photosynthesis rate is reached at lower light intensities. At low soil moisture stress and with little atmospheric

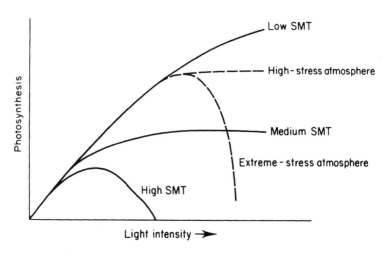

Figure 7.7 Expected effects of soil moisture stress (SMT)
and atmospheric stress of water upon photosynthesis at different light intensities.
(After Moss, 1965).

evaporative demand, photosynthesis continues to rise at
high light intensities.

Shade and cropping systems

In many multiple cropping systems "shade" is an essential
component.   Baldy (1963) reasoned that the many-layered
mixed communities traditionally grown in desert oases (e.g.
date palm + apricot + vegetables) may use water more efficiently in biomass production than pure stands because the
micro-climate may be favourably influenced by effects of
shading and windbreaking.   The upper storey creates a
favourable micro-climate for the storey below, and the crop
chosen for each successive lower storey is more mesophytic,
more shade tolerant and less light demanding than the
layers above.   Another example was given by workers at IRRI
(1974) who found that upland rice cannot only successfully
be grown under coconut trees, but that the rice may
actually benefit from the shade provided, especially in
areas with high radiation levels and droughts (Figure 7.8).
When water is a limiting factor, on cloudy days the plant's
stomata of both shaded and unshaded rice remain open.   For
the unshaded plants, however, on sunny days, the stomata
remain open only in the morning and close in the afternoon.
Hence, a large portion of the solar energy is wasted when
the crop is under water stress on sunny days.

Shelter and cropping systems

Shelter effects are best known from multiple cropping sys-

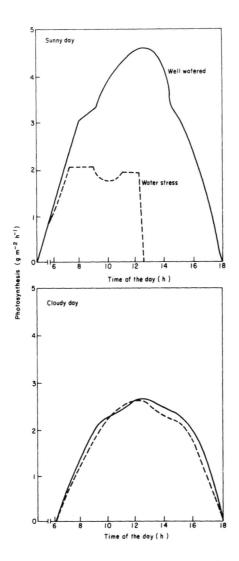

Figure 7.8   Photosynthesis on a sunny day and on a cloudy day
by a well watered crop (—) and a crop under water
stress (---).   (From IRRI, 1974).

tems with annual windbreaks.   The reason for the effects of
windbreaks can be classified in two categories:

(i)   the wind has a direct effect on plant growth
       and yield (e.g. winds may cause mechanical
       damage to the plants); and
(ii)   plant yields increase as an indirect effect
       of wind, through changes in the micro-
       climate, mainly on the lee side of the wind-
       break.

Windbreaks provide a mechanism to manipulate crop envi-

ronments in order to improve plant growth and water-use
efficiency. Windbreaks reduce the wind speed on the lee
side of a windbreak (see Figure 7.9) and evaporation from

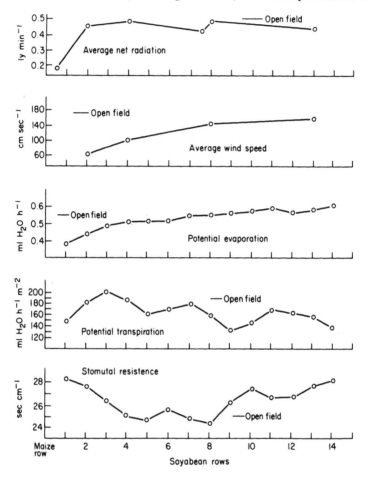

Figure 7.9   Some micro-meteorological and plant-water measurements as a function of soya bean row number, north
of a maize windbreak, together with measurements
in an open field for an August day in the corn belt
of the U.S.A. The day on which the measurements
were taken was characterized by water stress.
(After Radke and Hagstrom, 1970).

the soil. Hence, under dry conditions, the soil in the
sheltered area dries more slowly than soil exposed to drying winds. This slower evaporation and drying in sheltered
soil improves the conditions for seed germination (Rosenberg, 1966). Plants behind a windbreak respond to the
shelter with greater turgidity and wider stomatal aperture
in the leaves. Since wind speeds are slower in sheltered

areas, less water is required for transpiration. Further, when conditions become dryer, plants in open fields close their stomata more quickly than sheltered plants because their roots are unable to provide sufficient moisture for evaporation. This happens because plants under water stress react by partially closing their stomata until internal plant water deficits are relieved. The delay or avoidance of wilting in sheltered areas suggests that more efficient photosynthesis contributes to greater yields. Plants in sheltered areas grow taller and more water is normally consumed by these more vigorously growing plants. The fact that yields are generally greater in sheltered areas suggests that the protection leads to improved water-use efficiency.

Windbreaks are planted to modify the micro-climate. In many mixed and relay cropping systems, however, the micro-climate is also changed, sometimes inadvertently, by some action of one of the components of the association. This is illustrated by the "mulching acting" of the low statured component of a crop association. In this context mulching is defined as the application or creation of any soil cover that constitutes a barrier to the transfer of heat or vapor. Crop mulches reduce soil evaporation and temperature fluctuations in the soil. Hence, the soil micro-climate for the other crop in the association is changed, often beneficially.

PESTS AND DISEASES IN MULTIPLE CROPPING

Introduction

When considering the incidence of pests and diseases in multiple cropping systems, there are two widely contrasting possibilities:

- (i) Multiple cropping provides a longer period of plant life which is likely to increase insect and disease problems. More intensive cropping could change pest problems by creating a more favourable environment for pests and diseases by increased disturbance of the ecosystem; and
- (ii) Crop diversity may lead to greater pest stability and the longer period of plant life may allow naturally occurring biocontrol agents to sustain higher population levels. (Litsinger and Moody, 1976).

In multiple cropping systems, pests are a concern throughout the entire cropping period. The pests of the various crops do not only affect one crop. In sequential cropping systems the pests of one crop might be influenced by the previous crop while in mixed cropping systems, the pests of one crop might be influenced by the other component of the association.

97

For each species there is a range of pest and disease susceptibility which has to be taken into consideration in selecting the components of the cropping system. Generally, it is advantageous to combine pest or disease suscep- tible species with resistant species to reduce the absolute effects of the disease.

Crop rotation and diseases

The need for crop rotation is well known. Generally, the alteration of crop species decreases the incidence of pests and diseases. When making a crop rotation programme, crops should be selected which have the fewest pests in common. Crops which are botanically related have many pests and diseases in common and should not, therefore, be planted at the same time or in the same sequence. Crops which belong to the family of the Solanacea (i.e. tomatoes, potatoes and tobacco) are highly susceptible to nematodes, and many cultivars of these crops suffer from fungal diseases and insect pests.

Litsinger and Moody (1976) give examples of cropping pat- terns in Southeast Asia with crops which are botanically related and unrelated. (See Figure 7.10). In pattern I,

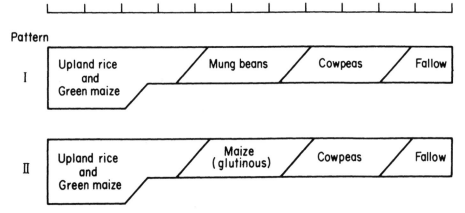

Figure 7.10 Two possible cropping patterns for Southeast Asia with botanically related and unrelated crops in rotation (After Litsinger and Moody, 1976).

pest problems would be expected from cowpeas following mung beans since they are both leguminous crops, attract similar pests and have similar growth habits. In pattern II, glu- tinous maize and cowpeas are not related, have different growth habits and attract different pests. In this pattern, however, pests could be transferred from the green maize to the glutinous maize. Since pest problems are normally greater with legumes than with maize, the second rotation is preferable.

From a pest ecology point of view, a rotation of a legume and a cereal usually has many advantages, and rotations such as wheat and soya beans, maize and common beans, and rice and mung beans are widely practised. Sequential cropping of maize can generally be practised without too many pest and disease problems and sequential cropping of rice is also widely practised. If a pest or disease breaks out, however, it is often necessary to introduce another crop. Heavy outbreaks of the brown planthopper (Nilaparvata lugens) has made continuous cropping of rice undesirable in some parts of Asia.

In contrast to maize and rice, crops such as tobacco and cotton should not be planted too frequently on the same land. Because pests and diseases seriously affect cotton production it should not be double cropped. Also, since cotton has many polyphagous insects, the other crops in a cotton rotation should be carefully selected.

Life cycles of pests and diseases are often synchronized with those of the host plants and are frequently determined by climatological and ecological conditions. Therefore, a pest can often only thrive when the host plant is in a certain stage of development. When the host plant moves to another stage of development, the pest often searches for another host. If there is no such host, the pest population decreases, or, if the period without an adequate host is long enough, disappears.

Crop arrangement, plant density and pests

Under natural conditions, the density of plant species is generally low since the species are grown in association with many other species. In a tropical rainforest, for example, more than 100 species per hectare are distributed more or less at random. This factor limits the incidence of monophagous pests.

On the other hand, under cultivation, plant densities of single species are much higher, and, in extreme cases, large areas are completely covered by one genotype with little genetic variability (e.g., large monoculture fields of maize or wheat of hybrid varieties planted at respectively 60,000 and 1 million plants per hectare). When a pest or disease breaks out at such densities, it spreads very rapidly. Because of this, it may be postulated that mixed cultures would experience slower rates of disease and pest transmission. Indeed, several authors have reported that there are fewer pest problems in mixed cropping than in sole cropping (Aiyer,1949; Batma, 1962; Trenbath, 1974).

When two or more species of which only one is a host to a certain pest or disease are planted in association, the presence of non-host plants acts as a barrier to the spread of the pathogen. IRRI (1975) found that the incidence of maize borer was significantly lower in maize/groundnut intercropping systems than in monoculture systems. The principal effect of intercropping was in reducing the reinfestation potential of the maize borer (See Table 7.1). The

Table 7.1
Effect of Row Spacing and Intercropping of
Maize on Maize Borer Incidence

Cropping System	Egg masses at 46 days after seeding		Pupal cases at 86 days after seeding	
	no. per 50 plants	no. per m^2	no. per 50 plants	no. per m^2
Maize in mixed culture at 20,000 pl./ha.				
maize + groundnuts	2	0.08	26	1.04
maize alone	8	0.32	32	1.28
Maize in mixed culture at 40,000 pl./ha.				
maize + groundnuts	5	0.40	37	2.96
maize alone	11	1.14	57	6.84
Maize alone at 60,000 pl./ha.	17	2.04	74	8.88

Source: IRRI Annual Report, 1975.

number of pupal cases per unit area was eight times greater when maize was grown as a sole crop than when intercropped with groundnuts. IRRI (1975) also found that the incidence of downy mildew was lower in maize/rice associations than in monoculture checks. Infestation was less at low maize populations (20,000 plants/ha intercropped with rice) than at high maize populations (30,000 plants/ha). It was concluded that the reduction could have played a major role in preventing an increase in the incidence of mildew over a large area.

Tall plants may hide short plants in an association and, hence, protect the adjacent host. An example of this can be found in Indonesia where tall and short early varieties of rice are grown in association in order to hide the short variety and prevent birds from devastating the crop.

Many pests and diseases thrive best under certain weather conditions. Sometimes, planting can only be done at a certain time of the year since otherwise the crops would be devastated. For example, at ICRISAT in India, delayed planting of sorghum increases shootfly risk. In Zimbabwe, wheat should not be planted before May 15th because during the warm weather preceding this date, the rust risk is much higher than when the crop is planted during cool weather. When the planting date of one crop in a multiple cropping system is governed by pests and diseases, the planting date of other crops in the system is also affected.

VIII Evaluation and productivity of different systems

INTRODUCTION

Evaluations of the productivity of cropping systems or of different crops should be done in quantitative terms. It is relatively easy to compare the productivity of crops and agricultural systems that produce similar products and use similar resources. If the product (e.g. crude protein, carbohydrates), and the resources used (e.g., fertilizer, land, tractor fuel) can be defined, the evaluation can be based on total production and the amount of resources used.

Before the productivity of a cropping system can be assessed, the basis upon which the yield will be measured must be decided. For monocultures, the most usual expression is some measure of weight per unit of land (e.g. kg/ha, lb/acre). In multiple cropping systems, however, because of the different species, direct comparisons cannot be made and the productivity can only be assessed using a common denominator.

The productivity of a multiple cropping system can best be evaluated using the yields of monocultures of the species in the system as the common denominator. When this approach is used, the monocultures must be grown at optimum plant densities since yields are a function of plant density. The optimum plant density, in turn, depends on agronomic and environmental conditions. Because associations of species can change the crop environment, an environment, which is sub-optimum for one species, must be exploited to the maximum degree if the crop is grown in association with one or more other species.

Multiple cropping evaluation demands techniques by which many types of crops and crop sequences can be tested under varying environments. Most statistical procedures developed for evaluating agricultural systems are primarily designed for monocultures. Because these procedures generally do not meet the requirements for the evaluation of multiple cropping systems, other methods have to be developed.

LAND EQUIVALENT RATIO

The Land Equivalent Ratio (LER) (Harwood, 1973) is used to

evaluate the productivity of mixed cropping systems. It is defined as: <u>the total land required using monocultures to give total production of the same crops equal to that of one hectare of mixed crop</u>. It is calculated by determining the ratio of the yield of a crop in a mixture with its yield in a monoculture. Figure 8.1 illustrates this

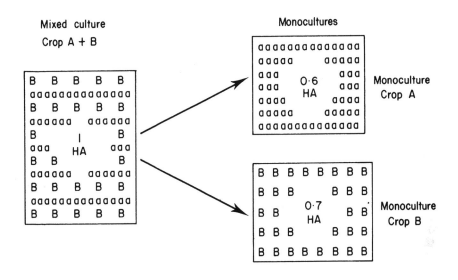

Figure 8.1   Diagrammatic presentation of the Land Equivalent Ratio concept.

concept. In this example, a cereal is mixed cropped with a legume. When optimum plant densities were used, the yield of the cereal in the mixed cropping system was 8.0 t/ha and the yield of the legume was 2.0 t/ha. The monoculture yields were 11.4 t/ha for the cereal and 3.3 t/ha for the legume. The ratios of multiple cropping to monoculture yields were thus 0.7 for the cereal and 0.6 for the legume. The Land Equivalent Ratio is defined as the sum of these two ratios, in this case 1.3. Total productivity is thus 30 per cent higher, or in other words, to produce the same amount of legume and cereal in monocultures, 30 per cent more land would be required.

The data presented in Figure 8.1 can also be represented as illustrated in Figure 8.2. The diagonal lines A-A and B-B labelled with percentage figures show the relative advantages in productivity over the monoculture check (Lines of equal LER) and the line 0-0 represents the base yield.

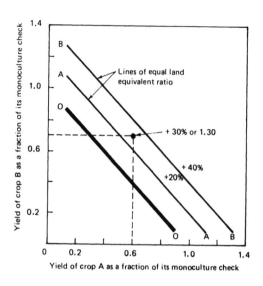

Figure 8.2   Relation between the yields of crop A and B given
in Figure 8.1.

In the example illustrated in Figure 8.2 only one mixed
culture was evaluated and compared with monocultures of the
crops in the association.   However, generally, comparisons
of several mixed cultures with each other are needed.   In
such cases comparisons with monocultures are the basis for
the evaluation. Application of these techniques are illustrated in Table 8.1 and Figure 8.3. The yield figures in
Table 8.1 were obtained from nine experimental plots.Seven
plots were assigned to mixed cultures of two species in
different proportions, and two to monocultures; one for
each species.   Some of the crop combinations were in the
form of a replacement series (see Chapter VI).The proportion of soya beans was lowest in association A (25 per
cent) and increased to 75 per cent in association G. Figure
8.3 illustrates the yields of Table 8.1 in graphical form.
The highest LER (1.40) was obtained for the associations
with the highest proportion of soya beans.

MULTIPLE CROPPING INDEX

The intensity of land-use can be expressed by the Multiple
Cropping Index which is calculated by dividing total crop
area by total cultivated land area and multiplying by 100
(Wang, 1975).

Table 8.1 Grain Yield of Maize and Soya Beans Grown together at different popula- tions and spatial arrangements and in Monocultures

Treatment	Maize Plant Population per Hectare	Soya Beans Plant Population per Hectare	Maize Yield (t/ha)	Soya Bean Yield (t/ha)	Maize Yield as Fraction of Monoculture A	Soya Beans Yield as Fraction of Monoculture B	Land Equivalent Ratio (A + B)
M a/	44.444	-	6,2		1,00		
A	33.333	83.333	5,5	0,3	0,88	0,23	1,11
B	29.620	74.074	4,9	0,4	0,79	0,30	1,09
C	30.476	76.190	4,7	0,5	0,75	0,38	1,13
D	29.630	148.148	4,5	0,5	0,72	0,38	1,10
E	22.222	166.666	4,5	0,6	0,72	0,46	1,18
F	26.666	266.666	4,3	0,7	0,69	0,53	1,22
G	11.111	249.999	2,1	1,4	0,33	1,07	1,40
S b/	-	333.333	-	1,3	-	1,00	-

a/ Maize monoculture check.
b/ Soya bean monoculture check.

105

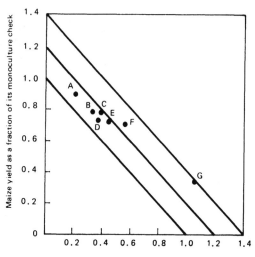

Figure 8.3 Relation between yields of maize and soya beans
in seven multiple cropping systems (From Beets,
1976).

$$Multiple\ Cropping\ Index\ (MCI) = \frac{Crop\ Area}{Cultivated\ Area}\ x\ 100\ per\ cent$$

The higher the multiple cropping index, the more crops are
planted and harvested from the same piece of land during
one year. This implies that both land and labour are more
intensively utilized and that some costs (e.g., soil
preparation, weeding) are lower per unit of output.

INPUT AND OUTPUT

In general terms, efficiency (E) can be described as an output (O) per unit of some input (I) (Spedding, 1973). Alge-
braically this can be represented as:

$$E = \frac{O}{I}$$

The output O may be measured in weight, money, energy or protein while the input I may be expressed in terms of land area used, energy, labour, fuel, fertilizer or any other resource utilized,including time. Time,land area and energy are normally important inputs in multiple cropping since they are scarce resources. Labour on the other hand is of less importance in measuring efficiency where family labour is used and where there are no alternative employment opportunities. Energy can be divided into "solar energy" used for photosynthesis and "added energy" (e.g., soil, electricity, farm machinery, fertilizers).

It is theoretically possible to compute the total energy used per unit of agricultural product and the energy value of the final crop product, and thus calculate the efficiency of a production system. By using a book-keeping approach a balance can be made of energy input and output. Although this approach has received a great deal of attention during the past decade, the methodology is still not fully developed and this approach cannot, as yet, be adopted as a standard method for evaluating cropping systems.

The above dealt with "added energy" and the energy contained in the harvested product. Another factor to consider is the efficiency of solar energy use. By measuring total photosynthesis per unit area of land it would theoretically be possible to deduce the "productivity" of a cropping system. Photosynthesis is closely related to leaf area and measurement of leaf area index, canopy cover and light transmission of canopies has, in some cases, been a valuable tool in assessing productivity. These methods, however, only help to explain differences and cannot be used as standard measures of productivity.

ENERGY AND PROTEIN PRODUCTION

Productivity can be assessed in terms of efficiency of energy and protein production per unit area of land per unit of time. It is sufficient to consider only energy and protein since these factors are of primary importance in most diets. Energy and protein must be considered separately since food crops contain both in different quantities and proportions. The balance between energy, protein and the constituent amino acids of the proteins must also be considered. The amino acids "lysine" and "methionine" are particularly important in tropical diets since lysine is often the major limiting amino acid in maize, which is a major staple crop, while methionine is the limiting amino acid in all sources of leaf protein.

There are several ways of measuring the energy and protein production of cropping systems. All depend on the use of the product (e.g. consumption by humans and animals of various kinds). The diagram in Figure 8.4 outlines the procedures which are generally followed. The yields of the

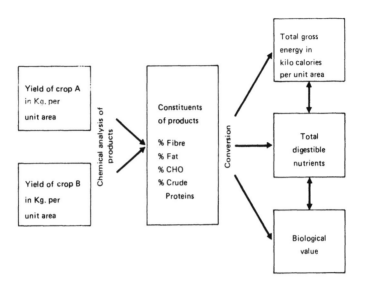

Figure 8.4   Diagrammatic presentation of a method of evalua-
             ting multiple cropping systems by converting
             yields to "energy".

crops are translated into their constituents which are then
summed and converted to energy. The Gross Energy does not,
however, necessarily represent the "Value" of the yield of
a cropping system. The quality of the proteins varies from
product to product and a combination of two or more
products in a particular proportion may have higher "biological value" than would be expected from the Gross Energy
Yield. This is illustrated in the histogram in Figure 8.5
which compares the yields of three mixed cropping systems
(two systems with maize and soya beans and one with maize
and groundnuts). In the maize/soya beans systems, on the
basis of mass or energy, the maize monocultures gave the
highest yields followed by the maize/soya mixed crop and
the soya monocultures. In terms of Fat (Ether extract),
Crude Protein and Methionine, the highest yield was given
by the maize/soya mixed crop. In terms of Lysine, one soya
monoculture check gave a higher yield than the correspon-
ding mixed culture. ·In another system (No. 2) the mixed
culture gave the greatest yield in terms of energy mass
crude protein and methionine. From the point of view of
Fat and Lysine, however, the groundnut monoculture provided
the highest yield with the maize/groundnut mixed crop a
close second.

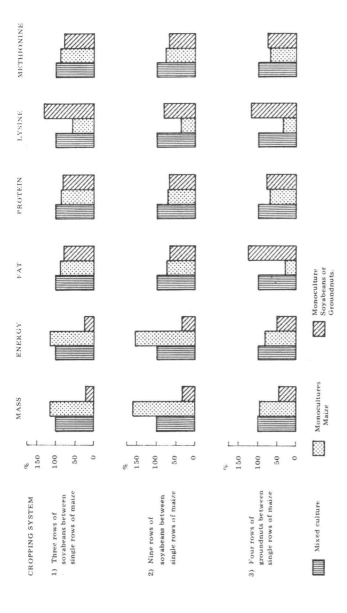

Figure 8.5 Comparison of the yields of three multiple cropping systems and their monoculture checks in terms of Mass, Energy, Fat, Protein, Lysine and Methionine (Beets, 1976).

Tarhalkar (1975) also found that mixed cropping systems provide produce of higher nutritive value. In particular he found that cereal legume mixtures contain proteins of superior nutritive value than monocultures as they usually supplement the deficient amino acids. Mixed cropping of sorghum with soya beans and groundnuts increased the Lysine yield up to 219 and 76 per cent respectively. This benefit of mixed cropping is of special importance in areas with protein deficient diets. Such areas exist in most developing countries and it is in these areas that multiple cropping often has greatest potential.

EVALUATION IN ECONOMIC TERMS

Although yields can be expressed in monetary terms, several difficulties are usually encountered with this approach. First, this method is only appropriate in areas where a cash economy exists. Second, the prices of produce and inputs often fluctuate seasonally and usually the ratio between them is not constant. The diagram in Figure 8.6 illustrates how the yields of a multiple cropping system can be compared with the yield of a monoculture. The multiculture consists of two crops (a cereal (A) and a legume(B)) and the monoculture with which the multiculture is compared can be either crop A or crop B. When this evaluation system is used, labour is considered as one of the "variable costs" and given a monetary value which can be done using "labour days" as an input and establishing a price for one labour day. In areas where labour is hired and paid wages, this is relatively simple. In cases where family labour is used or where hired labourers are paid in kind rather than cash, expressing the value of labour in monetary terms becomes difficult. This is the situation in most of the tropical farming systems.

In most instances output is expressed in terms of Gross Profits. If sufficient information is available on "overhead expenditures" (e.g. interest, capital repayment, depreciation), it is preferable to compare the Net Profits of the cropping systems being evaluated.

However, when using this approach there are several difficulties caused by seasonal price fluctuations. If the cereal yield is, say, four times the yield of the legume and the legume price is four times the cereal price, the gross income of the two monocultures would be equal. The gross income of the mixtures of the two crops would be directly related to the relative yield total (see also Chapter VI). As soon as the ratio of cereal and legume prices changes, the relative returns from the mixture having the greatest component of the crop which price is raised will be relatively more advantageous. Figure 8.7 illustrates the relative returns of a two phase replacement series at four price ratios and four levels of plant population. The monoculture yields of maize and soya beans and

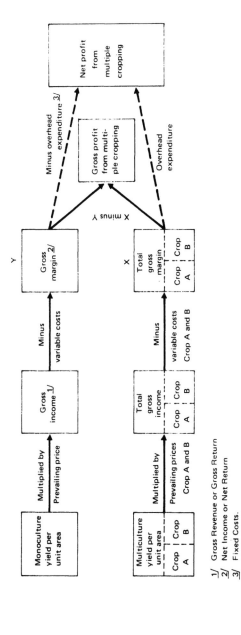

Figure 8.6 Diagrammatic presentation of an economic method of evaluating multiple cropping systems.

1/ Gross Revenue or Gross Return
2/ Net Income or Net Return
3/ Fixed Costs.

111

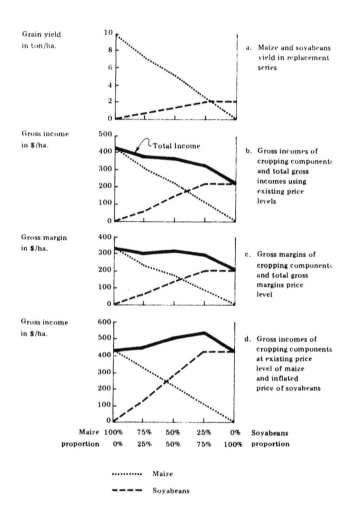

Figure 8.7 Economic evaluation of three mixed cropping systems
and two monoculture checks of maize and soya beans
at three different price levels.(From Beets, 1977).

for three mixed cultures in different proportions are
shown. Graphs b, c and d give the evaluation of the yields
of graph a in three different manners;namely,using the prevailing prices, with and without deducting the fertilizer
cost and graph d using an inflated price for soya beans.

These graphs clearly illustrate the sensitivity of the results of the economic evaluation to the assumptions made about prices. Given the volatile nature of prices and the difficulty in forecasting them accurately, this suggests that the results of economic evaluations must be viewed with caution.

IX Selection and design of suitable multiple cropping systems

THE ENVIRONMENT

Plant growth and performance in cropping systems are subject to environmental and management conditions. The environment is a function of all those factors related to land and climate, (e.g., topography, structure and texture of soil, rainfall and available moisture). Management is related to human activities (e.g., the method and time of planting, weeding). In this chapter, the environmental factors are considered first, then the management or human factors, and lastly the two are evaluated together in the design of cropping systems. Two illustrative examples have been used in this chapter, one from Africa (Salisbury, Zimbabwe) and one from Southeast Asia (Luzon, Philippines).

Prior to designing cropping systems in an area where farming is already practised, the existing cropping systems and the cropping environment must be understood and described. (See also Chapter X). Environmental classifications are normally based on rainfall, but in regions with a wide range of altitudes, they may also be based on temperature regimes. In other cases soil texture and topographic position are used as main parameters of the classification system. A useful start in environmental classification is a simple climatic diagram as shown in Figure 9.1. The basic data given in this diagram is by itself not sufficient for cropping systems design. From this elementary information, however, some broad conclusions can be drawn: although Nairobi (East Africa) is situated near the equator, temperatures are too low to support rice cultivation. Further, the site has a relatively long bi-modal rainfall and the annual precipitation is reasonably well distributed. From this information alone, however, it cannot be determined whether the climate is suitable for other crops (e.g., coffee, and pyrethrum). In order to draw such conclusions, more information is required (e.g., minimum and maximum day and night temperatures, duration of sunshine, and cloudiness).

The climatic diagram of Salisbury (Zimbabwe, Figure 9.2) is quite different from that of Nairobi and the following broad conclusions may be drawn: Although the altitude of Salisbury is lower than that of Nairobi (1,460 versus 1,820 m), a higher latitude makes the temperatures of Salisbury much lower for half the year. From May to October, tempe-

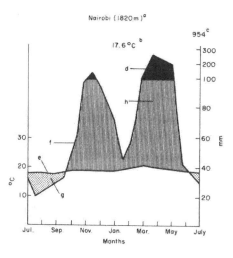

Nairobi (1820m)[a]

17.6 °C [b]

954[c]

Figure 9.1   Climate diagram of Nairobi, Kenya.
    Index: a) Altitude in meters.
             b) Mean annual temperature in °C.
             c) Mean annual precipitation in mm.
             d) For rainfall of greater than 100 mm
                 per month the scale is reduced to
                 1/10 (blackened).
             e) Curve for mean monthly temperature.
             f) Curve for mean monthly rainfall.
             g) Dry Season (dots).
             h) Rainy Season (lines).

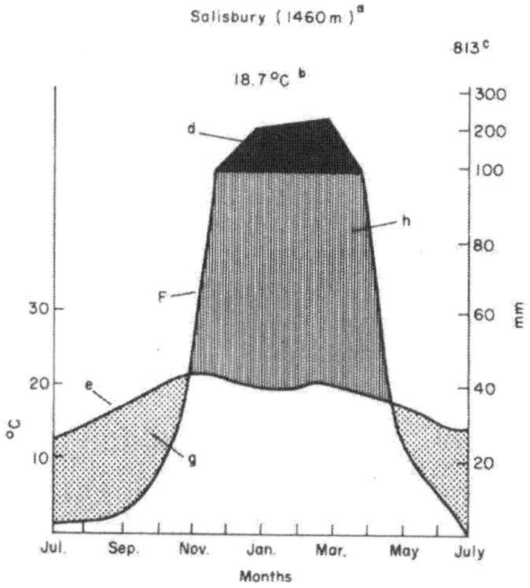

Salisbury (1460m)[a]

18.7 °C [b]

813[c]

Figure 9.2   Climate diagram of Salisbury, Zimbabwe.
             (For Index see Figure 9.1.)

ratures are too low for tropical and most sub-tropical crops. Temperatures and rainfall seem sufficiently high between November and April to support the growth of sub-tropical crops such as maize, millets and groundnuts. The dry season is so severe, however, that it is unlikely that any crop can be grown during this period. More information on factors determining the moisture balance (e.g., evaporation, dependability of rainfall) is, therefore, required to design cropping systems which make optimum use of the environment.

Rainfall pattern and moisture balance

Figure 9.3 gives more detailed information on Rainfall,

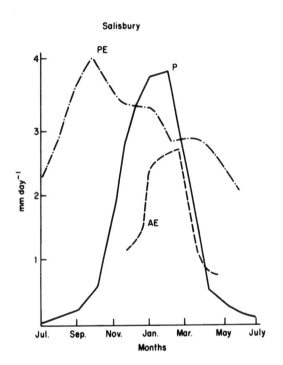

Figure 9.3 Rainfall (————), Potential Evapotranspiration (-.-.-.) and Actual Evapotranspiration (-----) for Salisbury. (After Donovan, 1961).

Potential and Actual Evapotranspiration of Salisbury. Because of the dry season, the agricultural potential (under rain-fed conditions) of the area depends largely on the length of the rainy season and the distribution of the rain during this period. The season lasts five to six months and for a period of at least 130 days the rainfall exceeds the Actual Evapotranspiration by a considerable amount. Since the growing periods of most crops do not exceed 130 days this suggests a satisfactory crop climate. Under rain-fed conditions, the cropping system should be designed so that the entire (moisture) season is fully utilized. This can be done by selecting late maturing crop varieties which have long growing periods and by planting them as early as possible. Under these circumstances, mixed cropping often better utilizes the season than monocultures.

For most Southeast Asian countries, rainfall data is usually available. Other macro-climatical parameters such as evaporation, temperatures, and relative humidities have, unfortunately, seldom been recorded for more than a decade. The agro-climatic classifications for evaluating cropping system potentials in Southeast Asia are, therefore, usually based on rainfall patterns or profiles. The International Rice Research Institute published a classification based on rainfall profiles in 1974 which recognizes eight climatic zones (See Figure 9.4). In this classification a wet month was defined as a month that receives over 200 mm rain and a pronounced dry season was defined as a period with at least two to three months with less than 100 mm rain per month. These amounts are based on two assumptions:

(i)   Losses due to evaporation are generally about 100 mm per month; and

(ii) Losses due to percolation and seepage are generally about 100 mm per month.

Another important environmental criterion is the number of consecutive wet months. If there are less than five consecutive wet months the potential for sequential cropping (under rain-fed conditions) is limited. Zone II of the classification, which includes areas with five to nine consecutive wet months, is of major interest for multiple cropping. In this zone, year-round rain-fed cropping is possible and because of the high rainfall during the height of the rains at least one crop will normally be rice grown under flooded conditions.

Radiation

In large areas of tropical America and Africa the agro-climate is determined by altitude, i.e., low mean annual temperatures associated with high altitudes. This is illustrated by Salisbury which is located approximately 18° south of the equator at an altitude of 1460 m. The mean monthly temperatures drop significantly after April. (See Figure 9.2). From May to August there is virtually no cloud cover which results in relatively high maximum day temperatures and low night temperatures. During July early morning frosts occur and double cropping of tropical crops is, therefore, not possible. It is, however, possible to

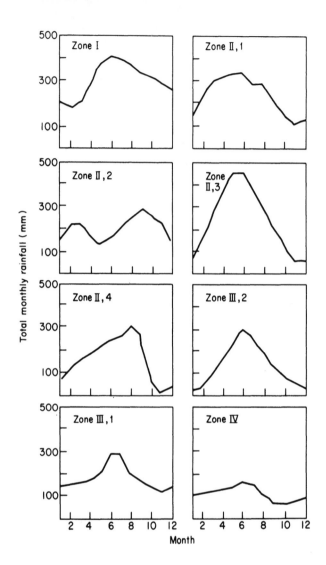

Figure 9.4   Southeast Asian climatic zones as differentiated
by rainfall pattern.   (After IRRI, 1974).

grow temperate crops during the colder weather.

In tropical high rainfall areas, cloud cover frequently limits yields during the rainy season and because of higher light levels, greater rice yields are usually obtained with irrigation in the dry season.

Radiation is not normally of great importance in the lowland tropics. Only when the cropping potential of an area within different altitudes is considered does radiation become important for cropping systems selection.

Cropping systems are closely associated with physiographic units, soil types, and soil fertility. On the basis of these parameters, zones can be defined as follows:

    (i)   Coastal plains;
    (ii) Sloping land;
  (iii) Hilly land;
   (iv) Rocky areas; and
    (v) Terraces.

Coastal plains are often used for lowland rice and estate crops such as sugarcane. Terraces are frequently used for vegetables and fruit trees. Rocky areas are normally unfit for cultivation except for tree crops and sloping areas are generally well suited to field crops such as maize and soya beans. When slopes are steep (over 15%) shifting cultivation is common. Irrigation is generally practised in coastal plains.

CROPPING SYSTEMS DESIGN AND PHYSICAL ENVIRONMENT

In Africa, where annual precipitation is over 600 mm, cropping systems are generally maize based. The information in Figures 9.2 and 9.3 suggests that maize is suitable to Salisbury. The shape of the rainfall curve indicates that the start of the season and the planting date of the crop depend on the start of the rains at the beginning of November. By the end of November moisture conditions are favourable as is shown in Figure 9.5.a. Moisture conditions become again less favourable when there is no rain for more than one week and mid-season droughts in January often reduce yields. When the rains tail off in March, growing conditions gradually deteriorate and an early stop of the season will reduce yields. By May the moisture balance is unfavourable for all plant growth although wheat or barley can be grown when irrigation is available. These crops are planted in mid-May and with complete irrigation the moisture balance is constantly favourable until the end of the growing season. "Early rains" may, however, hamper harvesting. Without irrigation, crops can only be grown from October to April, for example, relay cropping of maize and beans as illustrated in Pattern III. Maize is planted as early as possible and is widely spaced allowing beans to be relay-interplanted after the maize reaches maturity.

In tropical Asia, where precipitation is over 1,500 mm/annum with at least 200 mm/month rainfall for three consecutive months, cropping systems are generally based on rice. Zone II - 3 of Figure 9.4 can be used to illustrate which cropping systems are appropriate for this rainfall pattern. In this zone there are five to nine consecutive wet months and at least two months of less than 100 mm rain. This is, for example, found in parts of Central and East Java, southern Thailand, eastern and southeast Thailand

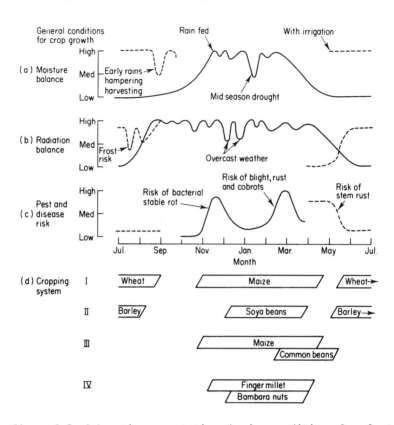

Figure 9.5 Schematic presentation showing conditions for plant growth taking into consideration: Moisture Balance, Radiation Balance, Level of Pest and Diseases risk. Cropping Systems that can fit the agro-climatological conditions of Salisbury, Zimbabwe are given under (d).

southern Burma and parts of the Philippines. Because low temperatures do not restrict plant growth, when the moisture balance is favourable, year-round cropping can be practised using a wide variety of crops. Since rice needs more water than other crops, and because it is the only crop that tolerates flooding, only rice is grown at the peak of the rains. Upland crops can be planted at the beginning and/or end of the rains to utilize residual moisture and higher light intensities during the dry season. (See Figure 9.6-I). Mixed cropping systems such as, for example, maize and groundnuts often best utilize the end of the rainy season. (See Figure 9.6-II).

System III shows a combination of a double and relay cropping system whereby transplanted rice is established as early as possible. The rice is followed by cowpeas using minimum tillage techniques and cucurbits are relay-planted later. One to two months of rain have to thoroughly soak the soil and some free water has to accumulate in order to

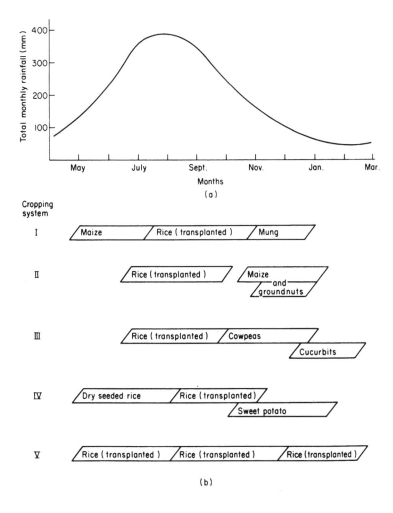

Figure 9.6   Rainfall pattern of Southeast Asian climatic zone
II-3 (after IRRI, 1974) and five cropping systems
that fit the rainfall pattern.

facilitate puddling of lowland rice soils. Dry seeding of
rice on unpuddled soil, on the other hand,allows establishment of rice with the first rains in May. Using early maturing varieties, a second crop of transplanted rice can be
grown and after maturity of the second rice crop the field
can be drained and planted with a crop like sweet potatoes
by using the residual moisture. This system is shown in
Figure 9.6-IV. System V represents a triple cropping system
of rice. This system might be preferable to a rotation of
rice and upland crops in areas with heavy soil and perennial irrigation as the soil does not have to be changed
from paddy to upland structure, a difficult process especially with heavy soils.

For intensive cropping there should generally either be an abundant supply of labour or a high level of mechanization. In sequential cropping systems, however, labour cannot always be substituted for mechanical power. One example of this is intensive rice-based cropping systems on heavy soils. With such systems land preparation requires considerable draft power and since the period available for land preparation is normally only two to four weeks, this operation often needs to be mechanized. Not only is the time available for land preparation for the first crop critical, but also the "turn-around period" is critical. In this context it should, however, be noted that minimum tillage techniques may be a suitable alternative for mechanized soil tillage.

The turn-around period for the double cropping systems of maize-wheat and soya-barley shown in Figure 9.5 is between two weeks and two months. Because this period is critical, many farmers prefer the soya bean system since this crop has a shorter growing period than maize and allows for a longer turn-around period. Although labour availability is not normally a limiting factor in Zimbabwe, farmers generally feel that double cropping systems are only possible at high levels of mechanization. They reason that good seedbed preparation using animal draft power or manual labour prior to the first rains is impossible. Land preparation is postponed until the first rains which postpones the planting of the first crop and shortens the turn-around periods of subsequent crops.

With the agricultural implements presently available, mixed cropping is difficult to mechanize. Because mixed cropping systems are labour intensive, they can only be practised in areas with an abundance of labour. Since mixed cropping has many important advantages, it appears desirable to either adapt existing farm implements to mixed cropping sytems or to design appropriate new machinery. With the widespread use of hydraulic devices and the introduction of electronics in farm mechanization, complex and versatile machines can now be designed. In future, it should be possible to develop machines capable of operating efficienly in mixed crop stands. Until then, mixed cropping will continue to be associated with low level technology farming requiring high labour inputs.

Irrigation

Irrigation and multiple cropping are closely associated, and according to Chao (1975), there is a significant statistical relationship between the two in Asia. Although irrigation facilitates multiple cropping, the relationship works both ways since multiple cropping may be necessary to justify investments in irrigation. The latter seems to be especially relevant in Africa. In many semi-arid areas of this continent, little or no multiple cropping is presently practised but construction of irrigation works would enable this. Since rivers are often far apart, large dams and canals have to be built and irrigation in these areas is, therefore, expensive. Investments are only justified if rates of return are high. Consequently, irrigation projects are only profitable with intensive year-round cropping. In the example given in Figure 9.5, no crops at all were grown during the dry season prior to irrigation. The construction of irrigation works was a capital-intensive venture, and the type of farming which followed the irrigation development had, therefore, to be commercial.

In the example for Asia given in Figure 9.6 the situation is quite different. Firstly, the rainfall pattern is much more favourable than in Salisbury and crops can be grown for most of the year without irrigation. Secondly, many areas in Asia are already partly irrigated and the introduction of irrigation is not normally as expensive as in Africa. When irrigation facilities are improved, cropping patterns are unlikely to change drastically; instead of one rice crop per year, more rice crops can be planted. Hence, the pattern will change from a main crop of rice during the peak of the rains followed by an upland crop to sequential rice cropping, possibly with one upland crop, although frequently continuous cropping of rice is preferred.

Partially irrigated areas in the tropics offer great potential for increasing multiple cropping. Other crops can be added to the main crop to make better use of available water in a growing season lengthened slightly by irrigation. In the example in Figure 9.5 the radiation balance is favourable for crop growth approximately three months before the moisture balance becomes favourable; (August against November). When crops such as late maturing maize and groundnuts are established with partial or supplementary irrigation one month prior to the start of the rainy season, the yields of these crops are generally 20 to 40 per cent higher than when planting is done under rain-fed conditions. When planting of these crops is done early, not only are yields higher, but since crops can be harvested earlier, turn-around periods are also shortened. This is a great advantage in the double cropping system of wheat and maize shown in Figure 9.5.d-I.

In the examples in Figure 9.6, the transplanted rice can always be planted "on time" when partial irrigation is available. Consequently, the main rice crop is harvested before the end of the rains and moisture conditions are likely to be more favourable for an upland crop following the rice.

CROPPING SYSTEMS DESIGN AND THE SOCIAL ENVIRONMENT

After technological changes necessary for the successful implementation of a new production system have been des-

123

cribed, the question of: "can society manage the degree of organization required for the implementation of the technological changes"?, arises. In communities where irrigation is already in use, it is possible to build on existing social structures. In areas where irrigation is unknown, a great deal of "social engineering" is required.

Cropping systems I and II of Figure 9.5, for example, require high levels of technology and organization. They involve full-scale irrigation, high yielding varieties and a high degree of mechanization. It is unlikely that these double cropping systems could be easily introduced into subsistence agriculture in Africa. On the other hand, the changes involved in switching from the presently widespread system II of Figure 9.6 to System V are less drastic and it would, therefore, be easier to introduce. Both systems involve rice, and growing rain-fed rice in a subsistence system is not very different from growing irrigated rice in a semi-commercial farming system. Further, mixed cropping is widespread in subsistence systems, and the change from, for example, a mixed cropping system of finger millet and bambara nuts (Figure 9.5.d-IV) to a relay cropping system of maize and common beans, with high yielding varieties, proper spacings and fertilizers (Figure 9.5.d-III) should encounter relatively few problems.

CROP HUSBANDRY

Tillage and land preparation

In most cropping systems, the time span during which crops can be planted to obtain maximum yield is quite short. This is particularly true with sequential cropping systems, and the success or failure of such systems often depends on the speed with which seedbed preparation can be carried out. Introduction of minimum and no-tillage techniques and use of chemical weed controls has facilitated multiple cropping systems in areas where they could not be used before on account of insufficient time for land preparation. At present, minimum and no-tillage techniques are successfully employed in many multiple cropping systems and yields are equal to or higher than those obtained with conventional tillage techniques. For example, Lewis and Phillips (1976) reported that no-tillage double cropped soya bean yields were generally equal or superior to yields of the same varieties grown by conventional methods in Kentucky, U.S.A.. Magbanua, et al (1977) found that many sequential cropping systems were successful with minimum or no-tillage in Asia.

The rainfall pattern usually has a pronounced effect on both the method of tillage and the speed of operations. Under arid conditions, soils are normally hard and difficult to work, especially in semi-arid areas with heavy soils. In Zimbabwe, for example, tillage is often delayed until the first rains soften the soil. Although extremely wet conditions usually hamper tillage, there is one important exception: land preparation for paddy. In fact, abundant rainfall often shortens the turn-around period for rice-based systems. (Magbanua, et al 1977).

A problem with tillage in multiple cropping systems is that operations are sometimes hindered by the presence of crop debris from a previous crop. This, however, also has the advantage that continuous soil cover improves the soil structure which, in turn, facilitates tillage.

Crops and varieties

The dietary requirements of the farm family are an important consideration for crop selection in subsistence farming systems. In Africa, for example, soya beans are unknown to most farmers and the cropping system proposed in Figure 9.5.d-I would, therefore, be difficult to introduce. On the other hand, soya beans are widely used in Asia and the Far East and the crop could be grown on both small subsistence and large-scale commercial farms.

In mixed cropping systems, there is usually a certain optimum proportion of the species in the mixture. This may, to some extent, be determined by dietary requirements (Tarhalkar, et al, 1975), economics or agronomic considerations. Andrews (1973) and Beets (1976) examined some agronomic factors and found that for groundnuts to be successfully grown in mixed stands, either very low population of the other crop or crops with a much longer growth cycle planted after the groundnuts are well established are required. In Zimbabwe, the system illustrated in Figure 9.5.d-I is preferred to system II for economic reasons, i.e., the gross incomes from maize + wheat are higher than those for barley + soya beans (System II).

In Asia and in some parts of West Africa, rice is a "prestige crop" and this crop is planted even in marginal conditions. On the other hand, in Eastern and Southern Africa maize is the preferred crop. Farmers' preferences, which are often traditional, also strongly influence crop selection.

The selection of appropriate varieties is important for multiple cropping systems. Use of early maturing varieties is often a prerequisite for multiple cropping (Frances, et al, 1975; Wang, et al, 1975; Tarhalkar et al, 1975) in order to grow as many crops as possible during the year. In Zimbabwe, for example, late maturing varieties of maize give higher yields than early maturing ones but, for cropping System I of Figure 9.5.d, a short maturing variety must be used since the turn-around period would be too short if a late maturing variety were used. In Southeast Asia, System IV of Figure 9.6, can only be used when the transplanted rice crop is early maturing. In this system, the second rice crop is planted in the middle of the rainy season rather than at the beginning as is normally the case. When late maturing rice varieties are used, the crop suffers from drought at the end of the rainy season.

When the Southeast Asian cropping systems of Figure 9.6 are considered the relevant characteristics of crops and varieties can be summarized as follows:

Maize: This crop can be grown for dry grain, green corn and sometimes for fodder. When the crop is grown for dry grain it should ripen during a period when the monthly rainfall is less than 100 mm. Maize varieties grown for green corn should provide soft, palatable cobs. For harvest during months with over 200 mm/month, green maize is preferable to dry grain maize because the harvest of green corn is not hindered by high rainfall. In view of this, the highest yields of dry grain maize can be obtained when the crop is planted in September-October while good yields of green maize require planting in May. Further, early October plantings can serve as fodder for cattle during the February to May dry season. Early maturing varieties are better than late maturing ones because they make the best use of the moisture season. Finally, varieties that are downy mildew tolerant are advantageous for May planting because of the higher downy mildew incidence in the rainy season. Short statured varieties with erect leaves are generally best for mixed cropping since they let more light through to the lower statured intercrop.

Rice: Dry seeded and upland rice can be more quickly established than transplanted lowland rice. The yield potential of rice grown under irrigated lowland conditions is, however, generally higher than dry seeded and upland rice. Further, rice can generally be planted under upland conditions only when the monthly rainfall is over 200 mm. Upland rice varieties should be deep rooted, as they will then be more tolerant to moisture stress. Since there are only few months that have over 200 mm of rainfall, early maturing upland rice varieties are preferable to late maturing ones. In view of this, all rice varieties give highest yields under rainfed conditions when they are planted with the first rains in April-May; June-July planting of upland rice is not desirable.

Mung beans: Mung beans, a typical upland crop, cannot be grown during the height of the rains since the crop is susceptible to water logging and cannot be harvested when wet. Therefore, this crop should be planted at the end of the rainy season, preferably around November. Late planting, however, increases the risk of moisture stress which reduces yields.

Groundnuts: Early maturing varieties generally have lower yield potential than late maturing varieties, but the former often fit better into cropping patterns. The plants generally tolerate heavy rains at the beginning of the growing period but after pod formation excessive rainfall induces diseases and physiological problems. To prevent sprouting and rotting in the field, the monthly rainfall should be less than 100 mm at harvest time. The crop tolerates some shading and is, therefore, often used in mixed cropping systems. In view of the above, groundnuts can best be planted around November.

Cowpeas: This is a versatile crop which can be grown for dry peas, can be harvested green as a vegetable and for fodder. It tolerates both heavy rains and drought but gives higher yields when grown under relatively high levels of light intensity during October-December.

Cucurbits: Because all cucurbits are sensitive to excessive moisture they are a typical dry season crop. Further, because initial growth is slow, this crop is well suited to intercropping and relay cropping systems. Planting is best done in November-December.

Sweet Potatoes: This crop tolerates heavy rain and over-shading during the first two months of its growing period. When the crop is relay-planted, the shade of a high statured crop is beneficial. After tubers have formed, excessive wetness leads to rotting. Because the crop is quite versatile in tolerating varying degrees of moisture it can best be planted from September to November.

There are significant interactions between cropping systems and varieties and different varieties are needed for different systems. When a crop is grown in a multiple cropping system, its yield potential may not be reached because an early maturing variety has to be planted in order to fit the time dimension of the entire system. Although the yield potential of individual crops may not be reached, the productivity of the combined crops in the systems is usually higher than the yield of monocultures with high yield potentials. Lastly, genetic improvement of crops in most countries has traditionally been concentrated on monocultures. When selecting varieties for multiple cropping, it is preferable to use a variety that has been found suitable to multiple cropping, or, a variety that has been specially bred for these systems.

Planting time and pattern

Because of changing environmental parameters, the yield potential of most crops varies with planting dates. Optimum planting times for crops in the Southeast Asian systems illustrated in Figure 9.6 range from one to three months.

In Figure 9.5, the situation is entirely different. The climate of Zimbabwe is much "harsher" than that of Southeast Asia and drought and low temperatures restrict the flexibility of planting period. Optimum planting periods for these systems are, therefore, shorter.

Unger and Stewart (1976), using a double cropping system of grain sorghum and sunflowers planted after wheat in the southern U.S.A., found that there is adequate time for initial seedbed preparation when the growing season is long. Planting times, however, are critical for the establishment of the second crop, since delays in planting may reduce yields through frost damage, failure of day-length-sensitive plants to mature, droughts, insects and diseases. Frost is of particular importance since the average harvesting time for irrigated winter wheat is late June, making the first of July a reasonable planting date for the second crop. A medium-early maturing grain such as sorghum, planted on the first of July would mature before frost in most years; late maturing varieties would not escape frost.

Hildebrand (1976) described an interesting multiple cropping system developed for El Salvador, shown in Figure 9.7 in which the use of double or twin rows of maize allowed the open space between the rows to be used for other crops without reducing the maize population. With irrigation, the system produced two full maize crops in one calendar year.

Soil type and fertility level

Soil conditions (e.g., structure, drainage, water holding capacity) influence cropping systems. For example, planting dates of all cropping systems of Figure 9.5 are influenced by the extent to which soil can be tilled. In Salisbury, the heavy soil is too hard to be tilled in September-October. On heavy soils high in montmorillonite content (see the example for Southeast Asia in Figure 9.6) lowland rice-based systems are the only feasible alternatives during the peak of the monsoon, since the soil can then not be tilled and maintained in an upland condition.

In both Salisbury, Zimbabwe and Luzon, Philippines the water storage capacity of the soils influences the length of the moisture season and thus crop selection for the later part of the season. When it is low, drought tolerant crops such as sorghum and cowpeas are preferable to maize and groundnuts.

In subsistence agriculture, where cash inputs are scarce, native soil fertility is an important determinant for cropping systems since year-round cropping, especially of cereal crops, requires high inputs of external nutrients. When fertilizers cannot be obtained, systems such as the double cropping of maize and wheat or barley and soya beans shown in Figure 9.5, or the double cropping systems of rice and maize and the triple cropping of rice as illustrated in Figure 9.6 cannot be used. On low fertility soils, the mixed cropping system of finger millet and bambara nuts of Figure 9.5.d will perform better than the high nutrient demanding relay cropping system of maize and common beans (System III).

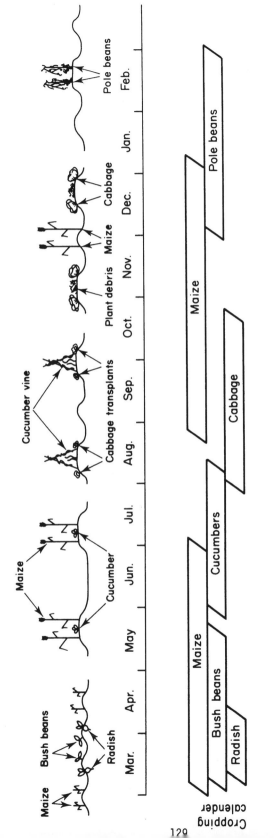

Figure 9.7 Diagrammatic presentation of a multiple cropping
system in El Salvador (After Mildebrand, 1976).

In relay and mixed cropping systems, plant interactions are significant and complex but to date, little scientific work has been done in this field and there is as yet no clear understanding of the nature of these interactions. Until more research is done, it seems advisable to fertilize associations which have a cereal as a major component on the basis of the requirements of the cereal with slight modifications and allowances for the other crop(s) in the association, for example, extra phosphate for leguminous crops.

Fertility management for sequential cropping should not differ significantly from sole cropping systems. Although a preceding crop in a sequential cropping system influences the fertilizer requirements of the following crop, interactions between crops are unlikely to be of greater significance than in sole crop rotations.

The double cropping system of maize and wheat of Figure 9.5 can only be practised when both crops receive relatively high applications of fertilizers. When the native nutrient status of the soil is medium, a maize crop yielding seven tonnes of dry grain per ha requires fertilizers providing the following amounts of major nutrients per hectare: 80-120 kgN; 45-70 kg P_2O_5; 35-55 kg K_2O. The requirements of wheat are similar or somewhat higher. The risk of applying high levels of nutrients to wheat is small since the crop is grown under complete irrigation and crop failures are, therefore, rare. The total fertilizer requirements of system II (soya beans and barley) are lower since soya beans require less nutrients. In Zimbabwe, soya beans grown in association with a cereal are sometimes not fertilized at all since the residual fertilizer meets the nutrient demands of the legume. System V in Figure 9.6, a triple cropping system of rice, has the highest fertilizer requirements of the systems given for Luzon. The requirements of the double cropping system of rice with the relay-planted sweet potato (System IV) also has high nutrient demands. These systems are only successful when nutrients can be supplied. In case of System V, complete irrigation during the dry season is also a prerequisite. System III, one rice crop with cowpeas and a relay planted crop of cucurbits, is best suited for low equilibrium farming since the cowpeas have a good nutrient uptake ability and reasonable yields under low levels of soil fertility. Cucurbits generally respond to fertilizer applications and chemical fertilizers can be successfully substituted for organic manures.

Pest and disease control

In sequential cropping, alternation of crop species is important - the sequence should be chosen so that the crops have the fewest pests in common. (See Figure 7.10). Therefore, cropping System II (double cropping of soya beans and barley) of Figure 9.5 is preferable to System I (maize and wheat). Furthermore, when crops are relay planted or planted in mixtures, it is advantageous to mix species of different botanical families to further reduce the potential damage from pests and diseases.

Harvesting and marketing

The speed at which crops can be harvested has important bearings on cropping systems. The double cropping systems illustrated in Figure 9.5, for example, can only be practised when crops are rapidly harvested since turn-around periods should be short. When harvesting crops grown in sequential cropping systems, crop residues should be left behind in an organized manner because irregular patterns may obstruct tillage and planting operations for the following crop.

Harvesting relay cropping systems can be difficult because the field is often not easily accessible and care should be taken that the remaining component of the system is not damaged. The spatial arrangement of crops in relay cropping systems should, therefore, allow for maximum accessibility. Consequently, twin rows are superior to equidistant spacing.

Harvesting mixed cropping systems has the same problems as relay cropping systems. If crops grown mature simultaneously, harvesting is facilitated. Associations of maize and groundnuts, maize and rice, sorghum and soya beans and finger millet are examples of systems in which the components can sometimes be harvested simultaneously. Pigeon peas, which have long growing periods, is an example of a crop which is normally harvested later than other crop(s) in the system.

Strip cropping systems are usually mechanized and the width of the strips should, therefore, facilitate machinery operation. A practical width of strips is a multiple of the width of the harvesting machine. An advantage of strip cropping a low statured crop with maize is that the maize lodges less than in pure stands. If the maize does lodge, however, harvesting of the low statured crop is more difficult than if this crop were grown in monocultures.

The efficiency of marketing systems is also an important factor. When products cannot be readily marketed, crops such as cassava, maize and sweet potato are advantageous. Harvesting can be delayed since these crops can be stored in the field. This, however, lengthens turn-around periods. When delivery to markets is slow, the cash flow of the farm is disrupted which may make the purchase of inputs for the following crops difficult.

SUMMARY

In the selection and design of a cropping system for a given region, the physical and human environments are the prime factors to consider. The most important factor influencing the physical environment is moisture balance. When the physical environment is described in quantitative terms and when the environmental requirements of crops are known, the two can be matched. The level of technology and the availability of economic resources are also important variables, particularly the level of mechanization and availability of irrigation.

Some cropping systems may require high levels of mechanization and can only be introduced when irrigation is available. When a cropping system can be practised from point of view of physical environment and level of technology attainable, agronomic management determines the success of the system. There is a significant interaction between tillage practises and cropping systems and in sequential cropping systems tillage can be a constraint.

Soil fertility is often a limiting factor in sequential cropping systems. While there are interactions between cropping systems and pest and disease management, generally this factor is not of dominant importance.

X Research

INTRODUCTION

Until recently, the western world, where most advanced scientific techniques were developed, paid little attention to research on multiple cropping systems, rather the focus was on monocultures. Similarly, cropping systems research in the tropics was concentrated on monocultures grown under relatively high levels of technology.

Indigenous cropping systems used by farmers in the tropics are frequently based on multiple cropping with low levels of technology. It has been increasingly recognized that such systems in the tropics are inherently different from western systems. These systems are often more superior in the environment in which they are practised and are frequently in balance with the technical (i.e. biological and physical) and human (i.e. social and economic) elements of the environment. Research on tropical cropping should address all elements of the systems and can, therefore, be based on a "systems approach" in which each cropping system is studied in toto.

CROPPING SYSTEMS RESEARCH

The objective of cropping systems research is to improve the use of a given quality and quantity of physical resources by increasing the efficiency of their utilization in crop production (Zandstra, 1977). The framework illustrated in Figure 10.1 outlines an interdisciplinary approach for cropping systems research based on the above objective.

A framework for cropping systems research

The first step in the design of new systems is assessing quantitatively the physical and biological environment (e.g. land appraisal studies, definition of ecological and climatic zones, soil surveys). On this basis, a number of potential cropping systems can be designed by drawing on the skills of soil scientists, agronomists and farm management specialists.

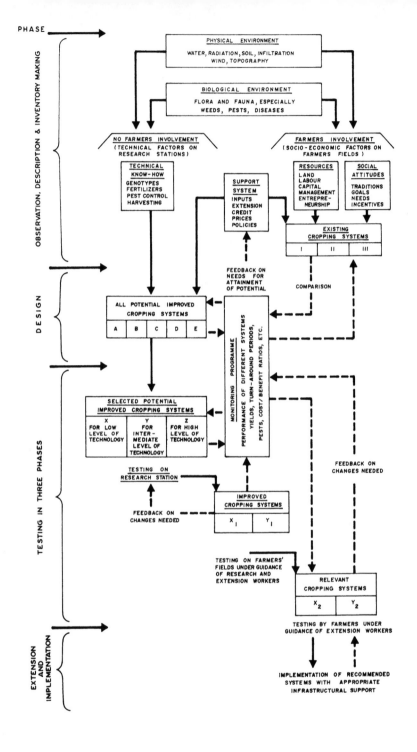

Figure 10.1 A framework for cropping systems research in the
tropics.

Concurrently, the farmer's present environment (e.g., current production methods, constraints) should be understood. The prevailing cropping systems should be described in a detailed and quantitative manner. This work can best be carried out by a multidisciplinary team of social scientists, extension specialists and agriculturalists. Many components of the support system (e.g., availability of credit, fertilizers, pesticides) can be described in quantitative terms. Data on the distribution of these components, availability and quality of extension services, product prices and other factors are usually difficult to express in quantitative terms. It is also important to assess the political environment since the efficiency of the infrastructural support system depends largely on political decisions. Assessing the present and future political environment is difficult, and erroneous judgements in this regard are often responsible for development failures.

The next major step is to compare potential with prevailing systems. Using the knowledge gathered from socio-economic conditions and the support system, the differences between potential cropping systems and those prevailing should be explained.

Using the knowledge thus obtained, the feasibility of the potential cropping systems should be evaluated, and those which seem most relevant should be tested under research station conditions. When the results of the testing are known, changes can be made, and the performance of the various systems tested should be compared with the existing cropping systems. At this stage, potential cropping systems can be classified according to the degree of technology needed and compared to the technology presently available to the farmer. Proposed cropping systems can only be successfully adopted when the level of technology required is in harmony with the level of technology presently in use.

When sufficient knowledge has been obtained about the cropping systems performance under controlled research station conditions, promising systems should be tested in an environment which resembles more closely practical farming conditions. This can be done on land rented from a farmer, with some assistance provided by the farmer, under the rigid control of research workers. Extension workers should also be involved at this stage. The penultimate step is to have the systems practised by farmers with the aid and guidance of extension workers. During this stage, good contact between farmers, extension officers and research workers is important.

The final step is to implement recommended systems with appropriate infrastructural support. When it is impossible to meet the necessary level of structural support, the improved system will probably fail under practical farming conditions. Since it is difficult to predict what level of support can be attained, it is advisable to consider several assumed levels of infrastructural support when designing the research programme. At higher levels, it can be assumed that the infrastructural support system is such that there is a possibility of substantially changing farming methods. If, at the observation phase, the infrastructural support system is poorly developed, more low-level technology systems should be designed. In absolute terms, low and intermediate level practices have potentially lower pay-offs than advanced level practices, but in terms of relevance, the first approach is less likely to fail.

Cropping systems research along these lines will only produce useful results if undertaken by a multidisciplinary research team. This does not necessarily mean that a large number of scientists have to be involved. One agriculturalist, for example, can participate in assessment of the natural resources, studying the existing cropping systems and planning and carrying out field trials. In general, versatile people are required, not necessarily highly specialized, but with extensive and broad experience. Scientists who meet this requirement can be permanently employed on the project and have the overall responsibility. If necessary, they can be assisted from time to time by specialists (e.g., soil scientists, plant breeders, entomologists, anthropologists).

Feedback and monitoring

It is important that new systems developed by agricultural researchers are capable of infinite repetition. That is to say, the essence of a viable system is that it does not harm the physical environment, for example, by allowing soil erosion. However, new systems cannot really be expected to be indefinitely applicable in exactly the same form because long-term unpredictable biological effects may occur, particularly if they are introduced over a wide area - through encouragement of new epidemic pests and diseases, for example. Further, the new system could have many side effects, e.g. it could induce changes in input and output prices, perhaps in social behavior through changing work patterns, or alterations in employer-employee relations.

Cropping systems research, therefore, needs to be a continuous process and feedback on necessary changes does not cease a few years after recommended systems have been implemented. To the contrary, feedback will have to be continuous and increases in importance as the biological and socio-economic environment in which the system is practised changes. Another important factor is that new technology always becomes available for testing. This, in turn, calls for further research, leading to further improved systems.

The test of relevance for a cropping system, therefore, goes far beyond adapting it to the physical environment alone. Adaptation to the human environment is equally important. The complex interactions between 'cropping systems performance' and 'total environment' can best be monitored by multidisciplinary research teams consisting of both technical and social scientists. However, comprehensive studies of all the parameters involved are not always feasible, and often an intuitive knowledge of interacting factors will have to suffice.

AGRONOMIC EXPERIMENTS WITH MIXED CROPPING SYSTEMS

Mixed cropping systems are the most intimate multiple cropping systems, and interference between crops is greatest. The physical and biological processes taking place in these systems usually also occur, to a lesser degree, in other multiple cropping systems.

Mixed cropping systems have high yield potential and are of interest from an agricultural research point of view. Since research on these systems is quite recent, an outline of some of the major aspects of mixed cropping experiments is discussed below.

Objectives

The main objective of trials with mixed cropping systems is generally to examine the benefits of certain crop mixtures over "other" cropping systems. Initially, these "other" cropping systems are monocultures of the species grown in the association. Later, when sufficient information is available on the advantages of crop mixtures over monoculture stands, different mixed cropping systems can be compared.

The objectives of such trials are:

    (i)   to compare the efficiency of several spatial arrangements of intercrop rows;

    (ii)   to investigate and develop a range of alternative intensive cropping patterns across the variability of water and soil conditions of the study area;

   (iii) to obtain insights into the degree of inter and intra-crop competition; and

   (iv) to compare resource-use, costs and returns of mixed cropping systems.

Variable treatments

For most mixed cropping trials a monoculture check is essential. To compare mixed cropping treatments with their monoculture check, it is sometimes necessary to have more than one monoculture check per trial. The monoculture checks should have certain plant populations and the sole stands should be grown at given levels of technology. In many of the experiments reported, mixtures have been achieved by adding together the plant populations used in the pure stand treatments (Agboola and Fayemi, 1971; Evans, 1960). One major disadvantage of this method is that the total population of the mixtures is then greater than that of the monoculture stands, which often means that plant populations in the monoculture stand are not sufficiently high to achieve maximum productivity. The mixtures may then give an apparent yield benefit simply because they are the only treatments which have total plant populations high enough to obtain maximum yield potential. Ideally the population pressure of all treatments in a mixed cropping trial should be equally high which can be achieved by planting the crops in a "replacement series". When using this approach, total population pressure remains constant and, at the same time, it is possible to create a range of mixtures with different proportions. Most of the pioneer work on competition done by de Wit and v.d. Bergh (1975) used the replacement series technique. Willey and Osiru (1972) used this technique when conducing mixed cropping trials with maize and beans and with sorghum and beans as did Beets (1976) with mixtures of maize and soya beans, maize and groundnuts and sorghum and soya beans. An example of treatments designed using this approach is given in Figure 5.4.

Experimental designs

Most statistical procedures developed for agricultural research are primarily meant for experiments involving sole crops. Mixed cropping and multiple cropping, however, requires the simultaneous testing and evaluation of several crops. Thus, techniques are required under which many crops and crop combinations can be tested under varying conditions. Three major issues arise:

(i)   because crop combinations have interactions among themselves, multi-factor experiments involving large numbers of treatments are advantageous;

(ii) because crops differ in their requirements (e.g., spacing, fertilizer level, weeding) and since they have to be planted together in one experiment, large experimental errors can be expected; and

(iii) because economic data are important for trial evaluation, the design of the trail should allow for the measurement of these data.

Workers at IRRI (1974) described a modified factorial design to test 63 multiple cropping systems (composed of three crops grown in pure stands, or intercropped at different durations of overlap, and under seven planting arrangements of maize) tested under three fertilizer levels and four weedcontrol levels. Although over 750 treatment combinations were possible only 256 were tested. Half of these were replicated while the rest were not. Trials of this size cannot often be carried out and, under most conditions, the number of levels of fertilizer application, weed control, etc. are restricted to one or two.

Randomized block designs are frequently used for the experiments. For example, Sooksathan and Harwood (1976) used this design to test three treatments of maize in pure stand, two treatments of rice/maize intercrop and a rice monoculture check. Dalal (1974) used the design to measure the effects of intercropping maize with pigeon peas and Beets (1977) used it to compare various maize/soya bean mixtures with monocultures of these crops. Other designs which are frequently used are the split-plot design (Liboon et al, 1975; Sooksathan and Harwood, 1976; Beets, 1976), and the factorial design (Evans, 1960).

Measurement of economic data is often important in reaching conclusions. Although this usually requires rather larger plot sizes than necessary for measurement of agronomic data replication is not usually necessary. Thus, more than one plot (without replication) could be used for the collection of economic, management and agronomic data and smaller plots (with replications) for detailed agronomic data.

AGRONOMIC RESEARCH ON OTHER MULTIPLE CROPPING SYSTEMS

The approach used for research on relay cropping systems hardly differs from that of mixed cropping. Relative plant populations, proportions, degree of inter- and intra-crop competition also play important roles.

Micro-climatological factors are generally important in strip cropping and annual windbreak systems. Research on such systems has frequently been concentrated on these aspects. Work done in this field by Radke and Hagstrom (1976) and Rosenberg (1973) is among the most advanced research work on multiple cropping systems. Economic aspects are important in sequential cropping systems, particularly the effects of interactions between cropping systems and turn-around periods.

SURVEY OF RESEARCH ON MULTIPLE CROPPING SYSTEMS

Until recently, limited systematic research has been carried out to improve or develop whole agricultural production systems which are adjusted to the specific agro-ecological and socio-economic conditions of regions or countries in the Tropics. At present, sufficient attention is rarely given to the working conditions of the small farmer. This, however, is slowly changing and interest in multiple cropping and whole agricultural production systems is increasing.

Although conditions in Asia, Africa and Latin America can vary considerably, there are also remarkable similarities in the working conditions on small farms, the performance of multiple cropping systems and their advantages over other methods of agricultural production. Therefore, a more international approach to research on multiple cropping systems could be rewarding. Because the character of this research involves multidisciplinary research teams consisting of rather large numbers of technical and social scientists, the major international centres are generally best equipped for this approach.

Although at present most of the research is carried out in Asia, in Africa, several institutes also conduct experiments while interest in the field is very recent in Latin America. The work being done in various countries and regions is summaried below:

Taiwan

Multiple cropping has been a special feature of agricultural research and development in Taiwan. For several decades comprehensive research has been conducted on many aspects of multiple cropping systems. Breeding and selection of varieties specially suited for these systems has been of particular interest and many excellent varieties have been developed.

The most important factors requiring further investigation were listed by ASPAC (1974) as follows:

(i) Adjustment of growing periods of rice to reduce typhoon damage: In Taiwan, double cropping of rice is important and more research will have to be carried out to determine optimum transplanting times, taking into consideration agro-ecological factors and the seasonal threat of typhoons;

(ii) Shortening growing periods: Although many varieties have already been specially bred for multiple cropping systems, much work in this field still remains to be done. Most cropping systems in the country are based on double cropping of rice and planting of non-rice crops must be arranged to fit between the two rice crops. Consequently, the breeding and introduction of varieties with short growing periods has become necessary. In many other parts of the world this is equally important; and

(iii) Adaptation to mechanization: Socio-economic conditions in Taiwan have changed quite drastically in recent decades; industrialization has been rapid and labour has gradually been withdrawn from the rural sector. Labour shortages are becoming an important factor in Taiwanese agriculture and multiple cropping systems which can be mechanized are now required.

The Philippines

An international multiple cropping research programme was initiated at the International Rice Research Institute in the Philippines in the 1960s. The cropping systems programme of the Institute employs the resource utilization approach to develop more efficient and productive cropping patterns for Southeast Asian rice farmers. Rice - based systems and mixed cropping research are the dominant areas of research.

Agronomic work includes studies on plant interrelationships, plant populations, and efficiency of light and fertilizer use. Pioneer work has also been done in the field of insect relationships and weed/crop interactions.

In more recent years, research on the technological components required for adequate management of cropping patterns has been undertaken. The programme has also examined the adaptation of cropping patterns to site variables such as soil, climate, landscape and the availability of labour and power. This involves large numbers of off-station experiments and is likely to increase the understanding of the factors that hinder agricultural development in the region.

The research areas which require furhter strenghtening include:

(i) An adequate description of the environmental factors influencing cropping systems performance;

(ii) A methodology to analyze and interpret the biological performance of cropping patterns as a function of the physical environment;

(iii) The development of cropping systems performance criteria;

(iv) Evaluation of component technology under different environmental conditions, with particular emphasis on the creation of a wide array of varietal alternatives, crop establishment methods and insect and weed management techniques;

(v) A clear understanding of the researcher-farmer test situation which is necessary to efficiently combine the farmer's experience and the researcher's expertise; and

(vi) A critical evaluation of the institutional requirements for cropping systems research.

Because only a small number of varieties are available which are particularly suited to mixed cropping, agronomic research on variety testing, breeding and selection will continue to play an important role. Some of the characteristics which are of special importance include shade, drought tolerance, short growing period and resistance to pests and diseases.

India

India has the second largest multiple-cropped area in the world (Mao, 1975) and there is considerable interest in research in this field. Many research papers have been published but few in-depth studies have been done. Basic research on the interrelationships between crops, the inter/intra competition ratio and the water-use efficiency of crop associations needs to be done. Although, on a country-wide basis, sequential cropping systems are likely to be able to make a greater contribution toward food production than mixed and relay cropping systems, Srivastava (1972) states that relatively little reseach has been done in this field.

The fields in which additional research is needed include:

(i)   Water household as a whole, particularly methods of adjusting cropping patterns to water regimes and vice-versa;

(ii)   Location-specific soil management techniques and practices in relation to cropping sequences;

(iii) The production-economics of various cropping sequences under differing agro-climatological conditions; and

(iv) The organizational structures, procedures and management practices required to meet the needs of the expanding agricultural technology.

ICRISAT: The farming systems research programme has been prominent in the operation of the International Crops Research Institute for the Semi Arid Tropics (ICRISAT) since it was established in 1973. The set-up of the programme is similar to that of IRRI. The programme covers a broad field, and is based on "production factor research" and "resource utilization research". The scope of agronomic research involves the following areas:

(i)   Relay and sequential cropping studies;
(ii)   Intercropping investigations;
(iii) Genotype evaluation trials;
(iv) Weed management systems; and
(v)   Methods to improve technology.

Considerable basic research is being carried out and the Institute has made a considerable amount of information available on multiple cropping systems for the semi-arid tropics.

In Africa research emphasis has been on monocultures rather than on mixed cropping systems. In 1934 Leakey stated that more priority should be given to research on indigenous methods of food production than on cash crops. Although, papers on mixed cropping by Evans (1960, 1962) and Grimes (1963) are now frequently quoted, at the time they were published, there was little interest and no follow-up. Willey and Osiru (1972) published excellent work on mixed cropping in Uganda but there has been no follow up on this work. Since 1964, socio-economic studies of traditional farming systems have been carried out at the Institute of Agricultural Research in Semaru, Nigeria including mixed cropping. (Norman, 1968, 1973, 1974; Baker, 1974; Andrews, 1972, 1975).

Sequential cropping systems have not played important roles in Africa. It seems possible that this is because in those areas where double cropping could be practised population pressures are generally quite low and the infrastructure and level of technology are not sufficiently developed. On the other hand, in the semi-arid and arid regions in Africa,where population pressures are relatively high and food scarcity common, sequential cropping can only be carried out when sophisticated irrigation systems are installed.

Tropical America

Until quite recently, limited systematic research on multiple cropping has been carried out in Latin America. Present interest is associated with the concern Latin American governments show for the small farmer (Pinchinat et al, 1976).

Working in Columbia, Gomez (1968) experimented with double cropping systems of maize and soya beans. Lepiz (1971) published data from mixed cropping trials of maize and beans, conducted in Mexico. Hildebrand and French (1974) did interesting work on an integrated system based on a combination of mixed and relay cropping in El Salvador and Flor and Frances (1975) worked on mixed cropping systems of maize and beans in Columbia.

More comprehensive studies of farming systems have been initiated by the Tropical Crops and Soil Department of Turrialba, Costa Rica. The research results indicate that under South American conditions, multiple cropping systems are generally more efficient than monocropping systems. These research efforts are, however, recent and consequently few of the results of experimental work have been implemented.

FUTURE RESEARCH NEEDS

The research information currently available indicates that multiple cropping systems are often superior to monoculture systems. The reasons for this are not yet fully understood because multiple cropping systems are complicated and to date only limited systematic research has been undertaken. Many agriculturalists now believe that improving the productivity of multiple cropping systems rather than attempting to replace them with capital- and energy-intensive monoculture technologies should be the research strategy for the future. During the past decade researchers have begun to follow farmers' innovations and adapting farmers' techniques rather than developing new techniques. This a positive trend.

There is uniform agreement that multiple cropping systems can only be completely understood when an interdisciplinary effort among scientists is used. This would not only include interaction among technical scientists (e.g., plant breeders, soil scientists, crop scientists) but also between this group and social and economic scientists. Only by using this approach can multiple cropping systems be fully understood and efficiently used to bring about a significant increase in world food production.

In particular the basic interrelationships between species planted in multiple cropping systems must be better understood. Basic studies are needed on root competition, how different species extract moisture and nutrients, and how the other components of the association are affected.

In view of the energy crisis and the high costs of chemical fertilizers, it is now extremely important that the whole question of the nitrogen cycle be better understood. Research needs to be undertaken on whether the nitrogen released by leguminuous crops can be taken up by a non-leguminuous species in an association. Answers to this and similar questions can only be found by using sophisticated and detailed research.

The technical research which is needed can be summarized as follows:

(i) determination of the inter/intra crop competition ratios for different crop associations;

(ii) selection of and breeding of crop varieties which are particularly suited to multiple cropping:
(a) improved architecture of some crops to reduce intercrop competition • (erect leaves, low stature, etc.);
(b) earlier maturing varietites;
(c) new cultivars which are adapted to different temperature and photoperiod conditions especially for sequential cropping systems;
(d) shade tolerant varieties for mixed and

relay cropping systems.

(iii) determination of the optimum time of planting for crops grown in multiple cropping systems in specific environments;

(iv) determination of optimum row spacings, absolute and relative plant population densities and level of weeding;

(v) understanding of the placement, timing, rates and allocations of chemical fertilizers more fully;

(vi) examination of multiple cropping systems as a whole and determination of the most efficient patterns for maximizing the utilization of solar radiation, water and nutrients;

(vii) measurement of differences in the microclimate induced by certain actions of components of crop associations and learning how to manipulate these to best advantage;

(viii) identification of the interactions between cropping systems and pest incidence; and

(ix) development of machinery appropriate for multiple cropping systems, particular for mixed cropping systems.

Advances in one discipline are likely to affect practices of other disciplines. In the same way changes of farmers' attitudes and practical farming will affect research. It is, therefore, important that communication between all parties involved in agricultural development, from policy makers to researchers to extension workers to farmers, be improved and that an interdisciplinary approach be used to the maximum extent possible when research on multiple cropping is undertaken.

References

Agboola, A.A. and Fayemi, A.A. (1971) *J. Agric. Sci. Camb.*, 77, pp. 219-225.

Aiyer, A.K.Y.N. (1949) *Indian J. Agric. Sci.*, 19, pp. 439-543.

Allen, L.H. (1955) *Agron. J.*, 66, pp. 41-47.

Allen, L.H. Jr., Lemayer, E.R. and Stewart, D.W. (1974) *Photosynthetica*, 8, pp. 84-207.

Andrews, D.J. (1971) IAR Report, *Samaru Project Paper*, p. 12, Samaru, Nigeria.

Andrews, D.J. (1972) "Intercropping with Sorghum in Nigeria", *Expl. Agric.*, 8, pp. 139-150.

Andrews, D.J. (1973) *Expl. Agric.*, 10, pp. 57-63.

Andrews, D.J. (1975) "Intercropping with Sorghum", *Samaru Research Bulletin 238*, Institute for Agricultural Research, Samaru, Nigeria, pp. 546-556.

Andrews, R.E. and Newman, G.I. (1970) *Oecol Plant*, 5, pp. 319-334.

Arnon, I. (1972) *Crop Production in Dry Regions*, Leonard Hill, London.

ASPAC (1974) "*Multiple Cropping Systems in Taiwan*". Food and Fertilizer Technology Center for the Asian and Pacific Region, Taipei, Taiwan.

Baker, E.F.I. (1974) "Research into Aspects of Farming Systems", *Samaru Misc. Papers*, Nigeria.

Baldy, C. (1963) *Ann. Agron.*, 14, pp. 489-534.

Barley, K.P. (1970) *Adv. Agron.*, 22, pp. 159-201.

Batra, H.N. (1962) *Indian Farming*, 11, pp. 17-19.

Beets, W.C. (1976) "Multiple Cropping Practices and their Fertilizer Requirements", FAO Fertilizer Programme, Paper presented at weekly staff meeting FAO, Bangkok, Thailand.

Beets, W.C. (1977) "Multiple Cropping", *World Crops*, Jan/Feb, pp. 25-27.

Beets, W.C. (1977) "The Agricultural Environment of Eastern and Southern Africa and Its Use", *Agric. Environm.*, 4, pp. 231-252.

Beets, W.C. (1977) "Multiple Cropping of Maize and Soya beans under a high level of Crop Management", Neth. J. Agric. Sc., 25, pp. 95-102.

Bergh, J.P. v.d. and Elberse, W.T. (1962) J. Ecol., 50, pp. 87-95.

Bergh, J.P. v.d. (1968) Versl. Landbouwk. Onderz. Ned. No. 714, pp. 1-71.

Bergh, J.P. v.d. (1975) Paper read at International Botanical Congress, Leningrad.

Benclove, E.K. (1970) "Crop Diversification in Malaysia", Incorporated Society of Planters, pp. 61-295.

Benclove, J.W. (1975) Cocoa Growing Under Rubber in Malaysia, Kuala Lumpur.

Birowo, A.J. (1975) The Philippine Economic Journal, 27, pp. 272-279.

Bowen, J.E. (1973) Plant Soil, 39, pp. 125-129.

Brady, N.C. (1974) The Nature and Properties of Soils, Mac Millan Publishing Co. Inc., New York.

Cable, D.R. (1969) Ecology, 50, pp. 27-38.

Chang, W. Chang, C.H. and F.W. Ho (1969) "Competition between sugarcane and intercrops for fertilizer tagged with P 32 and Rb 86". J. Agric. Assoc. China, 67, pp. 43-49.

Chao, C. (1975) "Improvements for increasing cropping intensity of paddy fields in Taiwan in the past five years", In Proceedings of the Cropping Systems Workshop, IRRI, Los Banos, Philippines.

Cocheme , J. (1968) "Agro-Climatological Methods", In Proceedings of the Reading Symposium, pp. 235-248, Unesco, Paris.

Cramer, P.J.S. (1957) "A review of literature of Coffee Research in Indonesia", Turrialba, Costa Rica. Interam. Inst. Agric. Sc. Misc. Publ. No. 12.

Cruz, A.F. and Alviar, N.G. (1975) "Multiple Cropping" J. Agric. Ec. and Develop., 21, pp. 14-22.

Daftardar, S.Y. and Savand, N.K. (1971) Plant and Soil, 34, pp. 303-307.

Dalal, R.C. (1974) "Effects of Intercropping Maize with Pigeon Peas on grain yield and nutrient uptake". Expl. Agric., 10, pp. 219-224.

Dalrymple, G.D. (1971) Survey of Multiple Cropping in less developed nations, U.S. Agency for International Development, Washington, D.C.

Donald, C.M. (1962) Aust. J. Agric. Res., 9, pp. 421-435.

Donald, C.M. (1963) Advan. Agron., 15, pp. 1-118.

Donovan, P.A. (1967) "Drought and Crop Ecology". In Proc. Symposium on "Drought and Development", Bulawayo, pp. 104-111.

Enui, D.P. (1972) Advan. Agron., 27, pp. 51-73.

Enyi, B.A.C. (1973) Expl. Agric., 9, pp. 83-90.

Evans, A.C. (1960) "Studies of Intercropping Maize or Sorghum with Groundnuts", East African Agric. and Forestry, 24, pp. 1-10.

Evans, A.C. and Sreedharan, A. (1961) East African Agric. and Forestry, 25, pp. 7-8.

Fisher, M.N. (1975) Tech. Comm. No. 15., Dept. of Crop Sci., Nairobi.

Frances, C.A. and Flor, C.A. (1975) "Adapting Varieties for intercropping systems in the Tropics", Paper presented in Symposium American Society Agronomy Knoxville, Tennessee, U.S.A.

Ganguli, B. (1930) Trends of Agriculture and Population in the Ganges Valley, Methuen, London.

Gomez, A.A. (1975) The Philippine Economic Journal, 27, pp. 288-295.

Gompertz, M. (1927) The Beginning of Agriculture, Gerald Howe, London.

Gray, R. (1953) Amer. J. Bot., 35, pp. 52-57.

Grimes, R.C. (1963) "Intercropping and alternate row cropping of cotton and maize" East African Agric. J., pp. 161-163.

Hao, D.P. (1972), Adv. Agronomy, 25, pp. 122-131.

Hacquart, A. (1944) "Project de culture mixte cacaoyers - heavea" Inst. Nat. pour L'Etude Agron. de Congo Belge, Publ. Serie, No. 28.

Hariot, T. (1888) A brief and True Report of the New Found Land of Virginia, Holbein Society Edin., Manchester.

Hart, R.D. (1975) "A bean, corn and manioc polyculture cropping system II. A comparison between the yield and economic return from monoculture and polyculture cropping systems", Turrialba, Costa Rica.

Harwood, R.R. (1973) Crop Interrelationships in Intensive Cropping Systems, IRRI Seminar, Philippines.

Harwood, R.R. and Price, E.C. (1976) In "Multiple Cropping" (M. Stelly, L. Eisele and J.H. Nauseef, eds), pp. 11-41, American Society of Agronomy, Madison, Wisconsin.

Herrera, W.A.T. and Harwood, R.R. (1973) "Crop Interrelationships in Intensive Cropping Systems", IRRI Saturday Seminar, July 21, 1973, Los Banos, Philippines.

Herrera, W.A.T. and Harwood, R.R. (1974) "Effect of time of overlap of corn in sweet potato under intermediate nitrogen levels", Paper read at the 6th Annual Convention of the Crop Science Society of the Philippines, Bacolod City, Philippines.

Hildebrand, P. and E.C. French (1974) "Un sistema Salvadoreno de multicultivos" Departemente de Economia Agricola, Centro Nac. de Tech. Agropecuaria, Ministerio de Agricultura y Ganaderia, El Salvador.

Hildebrand, P.E. (1976) In "Multiple Cropping". (M. Stelly, L. Eisele and J.H. Nauseef, eds), pp. 347-373. American Society of Agronomy, Madison, Wisconsin.

Hinson, K. (1975) "Nodulation Responses from Nitrogen applied to Soybean half-root". Agron. Journal, 67, pp. 799-803.

Huang, C.L. (1975) The Philippine Economic Journal, 27, pp. 126-138.

Hunter, J.R. (1961) Turrialba, 11, pp. 26-33.

ICRISAT (1976) Annual Report of the Farming Systems Research Program 1975-76, Hyderabad, India.

Indian Agricultural Research Institute (1972) "Research on Multiple Cropping", Ministry of Agriculture, New Delhi.

International Research Institute (1971) Annual Report 1971, Multiple Cropping, IRRI, Los Banos, Philippines.

International Rice Research Institute (1973) Annual Report 1973, Multiple Cropping, IRRI, Los Banos, Philippines.

International Rice Research Institute (1975) Annual Report 1975, Cropping Systems Program, IRRI, Los Banos, Philippines.

Johnston, A.A. (1968) "The Ford Foundation;s Involvement in Intensive Agricultural Development in India", New Delhi.

Kawano, K.H.G., Lucona, M. (1974) Crop Sci., 14, pp. 841-845.

Klute, A. and Peters, D.B. (1969) "Water Uptake and root growth", pp. 105-134, in W.J. Whittington (ed.), Root Growth, Butterworth, London.

Kranz, B.A., Kampen, J. and Associates (1976) "Informal Report of the Farming Systems research program of ICRISAT", Hyderabad, India.

Kung, P. (1971) "Irrigation Agronomy in Monsoon Asia," AGPC., Misc. FAO Rome p. 106.

Lakhani, D.A. (1976) Ph. D. Thesis, Reading University, England.

Lampeter, W. (1960) Wiss. Z. Karl-Marx Univ. Leipzig, Math. Naturwiss., 9, pp. 611-722.

Lang, A.L. (1949) Illinois Agr. Exp. Sta. Rept. AG 1437.

Lepiz, I.R. (1971) "Asociación de cultivos maíz-frijol", INIA, SAG, Mexico, Folleto Técnico No. 58.

Lewis, W.M. and Phillips, J.A. (1976) In "Multiple Cropping". (M. Stelly, L. Eisele and J.H. Nauseef eds), pp. 41-51. American Society of Agronomy, Madison, Wisconsin.

Liboon, S.R. and Harwood, R.R. (1975) "Nitrogen Response in Corn-Soybean Intercropping", Scrop Science Society of the Philippines, Bacolod City, Philippines.

Lin, C.F., T.S. Lee Wang, A.D. Chang, and C.Y. Cheng (1973) J. Taiwan Agric. Res., 22, pp. 241-262.

Litsinger, J.A. and Moody, K. (1976)In "Multiple Cropping", (M. Stelley, L. Eisele and J.H. Nauseef, eds), pp. 293-317. American Society of Agronomy, Madison, Wisconsin.

Magbanua, R.D., Roxas, N.M. Zandstra, H.G. (1977) "Comparison of the turn-around period of the different groups of patterns in Iloilo". Paper presented at the 8th Am. Meeting of the Crop Science Society of the Philippines, Benguet, May 5-7.

Manu, S. (1975) "Maximizing utilization of the rice areas in Chieng Mai Valley", pp. 126-143. In "Proceedings of the Cropping Systems Workshops", IRRI Los Banos, Philippines.

Mao, Y.K. (1975) The Philippine Economic Journal, 27, pp. 216-235.

Mather, J.R. (1974) Climatology, McGraw-Hill, New York, pp. 157-218.

Mittra, P.E. and Pande, D.A. (1972) in "Multiple Cropping". (Abulah Singh, ed), Indian Society of Agronomy, New Delhi, pp. 250-273.

Morales, J.O., Bangham, W.N. and Barrus, M.F. (1949) "Cultivos intercalados en plantaciones de Hevea", Inst. Interam. de Ciencia Agricolas, Boletin Tecnico No. 1.

Moss, D.N. (1965) Nature, 193, p. 587. Nelliat, E.V., Bavappa, K.V. and P.K.R. Nair (1974) World Crops, Nov., pp. 262-266.

Norman, D.W. (1973) Rural economics in the Zaria area, with general reference to agriculture, Samaru Research Bulletin 178, Institute for Agricultural Research, Samaru, Ahmadu Bello University, p. 9.

Norman, D.W. (1974) "Crop Mixtures under indigenous conditions in the Northern part of Nigeria", Samaru Bulletin 205, Zaria, Nigeria.

Norman, D.W. (1975) "Rationalizing Mixed Cropping under Indigenous conditions", The example of Northern Nigeria, Institute for Agricultural Research, Samaru, Ahmadu Bello University, Zaria, Nigeria, pp. 3-21.

Oshima, H.T. (1973) "Diversification and Development of Agriculture", In (H. Fukazawa Ed) Inst. of Develop. Econ., Tokyo.

Palada, M.C. and Harwood, R.R. (1974) "Relative return of Corn-rice Intercropping and Monoculture to Nitrogen Application", Crop Science Society of the Philippines, Naga City, Philippines.

Papadakis, J.S. (1941) J. Amer.,Soc.Agron.,33, pp. 504-511.

Paul, W.R.C. and Joachim, A.W.R. (1974) Trop. Agric., 90, pp. 257.

Pendleton, J.W., Belen, C.D. and Seif, R.D. (1963) "Alternating strips of corn and soybean versus solid plantings" Agron. J., 55, pp. 293-295.

Perkens, D.H. (1969) "Agricultural Development in China". Aldine, p. 45.

Phukan, P. (1970) In Dalrymple, p. 58.

Pinchinat, A.M., Soria, J. and R. Bazan (1976) In "Multiple Cropping" (M. Stelly, L. Eisele and J.H. Nauseef, eds.) pp. 51-63. American Soc. of Agron., Madison, Wisconsin.

Price, P. (1973) "The relative return of Corn-Rice Intercropping and Monocultures to Nitrogen application", IRRI, Los Banos, Philippines.

Radke, J.K. and Burrows, W.C.(1970). "Soybean plant response to temporary Field Windbreaks", Agron. J., 62, pp. 424-429.

Radke, J.K. and Hagstrom, R.T. (1970) Agronomy Journal, 66, pp. 273-278.

Rao, M.V. (1975) Cropping Systems in Southern India in "Proceedings of the Cropping System Workshop", IRRI, Los Banos, Philippines.

Reddi, G.H.S., Y.Y. Rao, and Y.P. Rao (1973) India J. Agric. Res., 7, pp. 177-187.

Revelle, R. and Thomas, H.A. (1970) "Population and Food in East Pakistan". Harward Center for Publication Studies.

Rosenberg, N.J. (1973) "Annual Windbreaks in crop production", Nature, 25, pp. 231-236.

Rosenberg, N.J. (1975) Micro-climate - The biological environment, John Willey & Sons, New York.

Samson, B.T. and Harwood, R.R. (1975) "The Effect of Plant Population/Density and Row arrangement on productivity of corn-rice intercrop", Crop Science Society of the Philippines, Bacolod City, Philippines.

Santhirasegaram, K. and Black, J.N. (1968) J. Br. Grassl. Soc., 23, pp. 234-239.

Sapier, O.L. de (1970) Agriculture and Diola Society, The Johns Muphings University Press.

Schwerdttfeger, F. (1954) All. Forstz., 9, pp. 278-282.

Singh, Y.D. and Kumar, K. (1972) In "Multiple Cropping" (Ambilah Singh ed), Indian Society of Agronomy, New Delhi, pp. 343-348.

Sivanappan, R.K., D. Chandrasekaran and E.S.A. Saifudeen (1976) "Skip-furrow Irrigation for Cotton". Indian Farming, June 1976, pp. 11-15.

Sooksathan, and Harwood, R.R. (1976) "A comparative Growth analysis of intercrop and monoculture plantings of rice and corn", IRRI Saturday Seminar, Feb. 21, 1976.

Spedding, C.R.W. (1973) The Biology of Agricultural Systems Academic Press, London.

Srivastava, H.P. (1972) In "Multiple Cropping" (Ambilah Singh (ed) Indian Society of Agronomy, New Delhi.

Stern, W.R. and Donald, C.M. (1962) Aust. J. Agric. Res., 13, pp. 615-623.

Syarifuddin A.E. Ismail and J.L. MacIntosh (1973) Contr. Res. Centr. Agric. Bogor, No. 12, p. 13.

Tarhalkar, P.P. (1975) "Changing Concepts and Practices of Cropping systems", Indian Farming, 25, pp. 3-7.

Trenbath, B.R. (1970) J. Appl. Ecol., 12, pp. 189-200.

Trenbath, B.R. (1974) Adv. Agron., 26, pp. 177-210.

Trenbath, B.R. (1975) J. Appl. Ecol., 12, pp. 189-200.

Unger, P.W. and B.A. Stewart (1976) In Multiple Cropping (M. Stelly, L. Eisele and J.H. Nauseef, eds.) pp. 255-273. American Society of Agronomy, Madison, Wisconsin.

Vidal, P. (1965) "Influence de l'Acacia albida sur le sol", IRAT, Senegal.

Vigo, A.H.S. (1965) "A Survey of agricultural credit in the Northern Region of Nigeria", Mimeographed (unpublished).

Wagner, L. and Mikesell, M.V. (1965) Readings in Agricultural Geography, The University of Chicago Press, Chicago.

Wallace, R. (1887) India in 1887, Edinburgh.

Wang, Y. and Yu, Y.H. (1975) The Philippine Economic J.,27, pp. 26-47.

Wang, P.E. (1975) The Philippine Economic J., 27, pp. 26.

Whittlesey, D. (1974) In Readings in Agricultural Geography (Wagner, P.L. and Mikesell, M.W. Eds.), The University of Chicago Press, Chicago.

Willis, J.C. (1914) Agriculture in the Tropics,Univ. Press, Cambridge.

Willey, R.W. and Osiru, D.S.O. (1972) J. Agric. Sci. Camb., 79, pp. 517-529.

Willey, R.W. and Rao, M.R. Oyen, L. (1976) "Suggested Research priorities in cropping systems", ICRISAT, Hyderabad, India (unpublished).

Wischmeier, W.H. and Smith D.W. (1965) "Predicting rainfall erosion losses from crop land", Agric. Handbook 282, US Depart. Agric.

Zandstra, H.G. (1977) "Crop Intensification for the Asian rice farmer", IRRI Paper, Los Banos, Philippines.

Index